ENERGY SCIENCE, ENGINEERING AND TECHNOLOGY

CYBERSECURITY FOR THE ELECTRIC SMART GRID

ELEMENTS AND CONSIDERATIONS

ENERGY SCIENCE, ENGINEERING AND TECHNOLOGY

Additional books in this series can be found on Nova's website under the Series tab.

Additional E-books in this series can be found on Nova's website under the E-books tab.

ENERGY POLICIES, POLITICS AND PRICES

Additional books in this series can be found on Nova's website under the Series tab.

Additional E-books in this series can be found on Nova's website under the E-books tab.

ENERGY SCIENCE, ENGINEERING AND TECHNOLOGY

CYBERSECURITY FOR THE ELECTRIC SMART GRID

ELEMENTS AND CONSIDERATIONS

PRESTON BARKER
AND
RUDY F. PRICE
EDITORS

Nova Science Publishers, Inc.
New York

For permission to use material from this book please contact us:
Telephone 631-231-7269; Fax 631-231-8175
Web Site: http://www.novapublishers.com

Library of Congress Cataloging-in-Publication Data

Cybersecurity for the electric smart grid : elements / editors, Preston Barker and Rudy F. Price.
p. cm.
Includes index.
ISBN 978-1-61942-353-4 (hardcover)
1. Smart power grids--Security measures--United States. 2. Computer networks--Security measures. I. Barker, Preston. II. Price, Rudy F.
TK3105.C93 2011
621.319'2028558--dc23
2011045828

Published by Nova Science Publishers, Inc. † New York

CONTENTS

PREFACE

The "Smart Grid" is the name given to the evolving electric power network as new information technology systems and capabilities are incorporated. While these new components may add to the ability to control power flows and enhance the efficiency of grid operations, they also potentially increase the susceptibility of the grid to cyber attack since they are built around microprocessor devices whose basic functions are controlled by software programming. This book examines the "Smart Grid" where the potential for a major disruption or widespread damage to the nation's power system from a large scale cyberattack has increased focus on cybersecurity.

Chapter 1 - Electricity is vital to the commerce and daily functioning of United States. The modernization of the grid to accommodate today's uses is leading to the incorporation of information processing capabilities for power system controls and operations monitoring. The "Smart Grid" is the name given to the evolving electric power network as new information technology systems and capabilities are incorporated. While these new components may add to the ability to control power flows and enhance the efficiency of grid operations, they also potentially increase the susceptibility of the grid to cyber (i.e., computer-related) attack since they are built around microprocessor devices whose basic functions are controlled by software programming. The potential for a major disruption or widespread damage to the nation's power system from a large scale cyberattack has increased focus on the cybersecurity of the Smart Grid.

Chapter 2 - The electric industry is increasingly incorporating information technology (IT) systems into its operations as part of nationwide efforts— commonly referred to as smart grid— to improve reliability and efficiency. There is concern that if these efforts are not implemented securely, the electric grid could become more vulnerable to attacks and loss of services. To address this concern, the Energy Independence and Security Act of 2007 (EISA) provided the National Institute of Standards and Technology (NIST) and Federal Energy Regulatory Commission (FERC) with responsibilities related to coordinating the development and adoption of smart grid guidelines and standards.

Chapter 3 - Title XIII of the Energy Independence and Security Act of 2007 (EISA) states, "It is the policy of the United States to support the modernization of the Nation's electricity transmission and distribution system to maintain a reliable and secure electricity infrastructure." The protection and resilience of critical national infrastructures is a shared responsibility of the private sector, government, communities, and individuals. As the complexity, scale, and interconnectedness of today's infrastructures have increased, it has changed the way services and products are delivered, as well as the traditional roles of owners, operators, regulators, vendors, and customers.

Chapter 4 - My testimony summarizes the Commission's oversight of the reliability of the electric grid under section 215 of the Federal Power Act (FPA) and the Commission's implementation of that authority with respect to cyber security primarily through Order No. 706. I also will describe some of the current limitations in Federal authority to protect the grid against physical and cyber security threats, and also comment on the cyber security discussion draft. The Commission currently does not have sufficient authority to require effective protection of the grid against cyber or physical attacks. If adequate protection is to be provided, legislation is needed and my testimony discusses the key elements that should be included in legislation in this area.

Chapter 5 - NERC's mission is to ensure the reliability of the bulk power system of North America and promote reliability excellence. NERC was founded in 1968 to develop voluntary standards for the owners and operators of the bulk power system. NERC is an independent corporation whose membership includes large and small electricity consumers, government representatives, municipalities, cooperatives, independent power producers, investor-owned utilities, independent transmission system operators and federal power marketing agencies such as TVA and Bonneville Power Administration.

In: Cybersecurity for the Electric Smart Grid
Editors: P. Barker and R. F. Price
ISBN: 978-1-61942-353-4

Chapter 1

THE SMART GRID AND CYBERSECURITY - REGULATORY POLICY AND ISSUES*

Richard J. Campbell

SUMMARY

Electricity is vital to the commerce and daily functioning of United States. The modernization of the grid to accommodate today's uses is leading to the incorporation of information processing capabilities for power system controls and operations monitoring. The "Smart Grid" is the name given to the evolving electric power network as new information technology systems and capabilities are incorporated. While these new components may add to the ability to control power flows and enhance the efficiency of grid operations, they also potentially increase the susceptibility of the grid to cyber (i.e., computer-related) attack since they are built around microprocessor devices whose basic functions are controlled by software programming. The potential for a major disruption or widespread damage to the nation's power system from a large scale cyberattack has increased focus on the cybersecurity of the Smart Grid.

Federal efforts to enhance the cybersecurity of the electrical grid were emphasized with the recognition of cybersecurity as a critical issue for electric utilities in developing the Smart Grid. The Federal Energy Regulatory Commission (FERC) received primary responsibility for the reliability of the

* This is an edited, reformatted and augmented version of a Congressional Research Service publication, CRS Report for Congress R41886, from www.crs.gov, dated June 15, 2011.

bulk power system from the Energy Policy Act of 2005. FERC subsequently designated the North American Electric Reliability Corporation (NERC) as the "Electric Reliability Organization" (ERO) with the responsibility of establishing and enforcing reliability standards. Compliance with reliability standards for electric utilities thus changed from a voluntary, peer-driven undertaking to a mandatory function. The Energy Independence and Security Act of 2007 (EISA) later added requirements for "a reliable and secure electricity infrastructure" with regard to Smart Grid development. NERC is also responsible for standards for critical infrastructure protection (CIP) which focus on planning and procedures for the physical security of the grid. Self-determination is a key part of the CIP reliability process. Utilities are allowed to self-identify what they see as "critical assets" under NERC regulations. Only "critical cyber assets" (i.e., as essential to the reliable operation of critical assets) are subject to CIP standards. FERC has directed NERC to revise the standards so that some oversight of the identification process for critical cyber assets was provided, but any revision is again subject to stakeholder approval. While reliability standards are mandatory, the ERO process for developing regulations is somewhat unusual in that the regulations are essentially being established by the entities who are being regulated. This may potentially be a conflict of interest, especially when cost of compliance is a concern, and acceptable standards may conceivably result from the option with the lowest costs. Since utility systems are interconnected in many ways, the system with the least protected network potentially provides the weakest point of access.

Cybersecurity threats represent a constantly moving and increasing target for mitigation activities and mitigation efforts could likewise spiral upward in costs. Recovery of costs may present a major challenge especially to distribution utilities and state commissions charged with overseeing utility costs. EISA only requires states to consider recovery of costs related to Smart Grid systems. FERC has jurisdiction over the bulk power grid, and cannot compel entities involved in distribution to comply with its regulations. Recoverability from a cyber attack on the scale of something which could take down a significant portion of the grid will likely be very difficult, but maintaining a ready inventory of critical spare parts in close proximity to key installations could quicken recovery efforts from some types of attack. The electricity grid is connected to (and largely dependent on) the natural gas pipeline, water supply, and telecommunications systems. Technologies being developed for use by the Smart Grid could also be used by these industries. Consideration could be given to applying similar control system device and system safeguards to these other critical utility systems.

INTRODUCTION

Electricity is vital to the commerce and daily functioning of United States. The electrical grid of the United States consists of all the power plants generating electricity, together with the transmission and distribution lines and their associated transformers and substations which bring power to end-use customers. Electric power generation in the United States is currently dominated by the use of combustible fuels, such as coal and natural gas, or from biomass. These fuels are burned to produce steam in boilers which is used to turn turbine-generators which produce electricity. Nuclear power uses heat from the decay of radioactive elements to produce steam. However, electricity can also be generated directly by wind turbines, solar power, geothermal energy, and hydropower. Figure 1 illustrates the phases of the electric power production process. Electricity from large central station power plants is routed to a step-up transformer which raises the voltage so the power can be sent over high-voltage transmission lines to step-down transformers in substations which then transfer this energy at lower voltages via distribution lines to industrial, commercial, and residential users.

Simplified Schematic

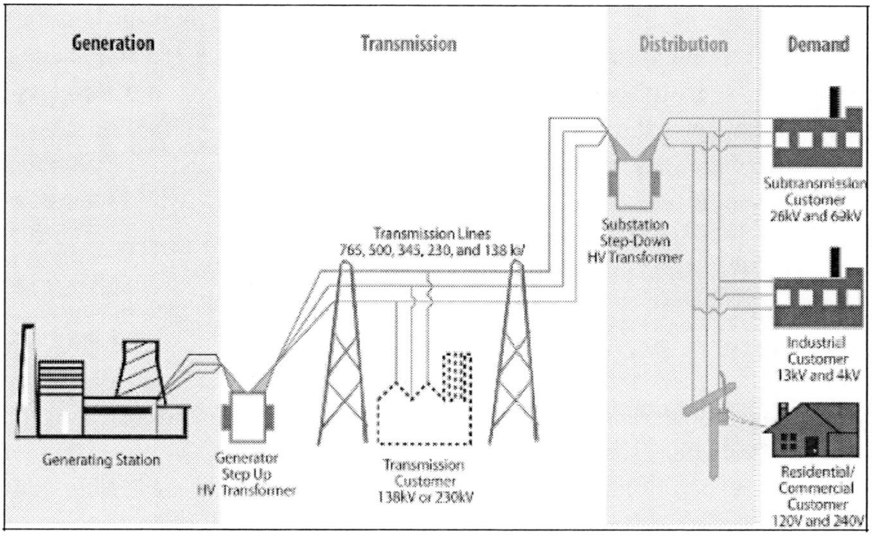

Source: CRS, based on graphic found at https://reports.energy (p. 5).

Figure 1. Elements of the Electric Power System.

Generally, electricity must be used as soon as it is produced because large amounts of electricity cannot be easily stored.

Originally, the individual company systems were not linked, but with greater electricity demand came the necessity of sharing generation resources. This sharing of generation resources required an interconnection of separate company systems to enable power sales and transfers. These aggregated power systems form three major "interconnections"—the Eastern and Western interconnections, and the Electric Reliability Council of Texas which includes most of the state of Texas. Within these interconnections are reliability regions, and a number of balancing authorities[1] which "integrate resource plans ahead of time, maintain load-interchange generation balance" within a balancing authority area, and "support interconnection frequency in real-time."[2] The grid also connects the many publicly and privately owned electric utility and power companies in different states and regions of the United States[3] (and in Canada and Mexico).

Congress passed the Public Utility Regulatory Policies Act of 1978 (P.L. 95-617) which allowed non-utility power producers into wholesale power markets (e.g., for the sale of electricity to entities other than the end-user of power). The passage of the Energy Policy Act of 1992 (P.L. 102-486) served to further promote greater competition in the bulk power markets.[4] As a result, in many parts of the United States, the electric power industry began to transition from highly regulated, local monopoly companies which generated, transmitted, and distributed electricity to end-use customers, to a business in which power generation is competitive while the industry's transmission and distribution functions are still highly regulated. While the federal government regulates the electric power transmission function and wholesale power markets, regulation of the distribution function of the electric power business is still largely carried out by state government agencies.

Much of the infrastructure which serves the U.S. power grid is aging. The average age of power plants is now over 30 years; most of these facilities were originally designed to last 40 years.[5] Electric transmission and distribution system components are similarly aging, with power transformers averaging over 40 years of age,[6] and 70% of transmission lines being 25 years or older.[7] As components of the system are retired, they are replaced with newer components often linked to communications or automated systems.

With changes in federal law,[8] regulatory changes, and the aging of the electric power infrastructure as drivers, the grid is changing from a largely patchwork system built to serve the needs of individual electric utility companies to essentially a national interconnected system capable of accommodating massive transfers of electrical energy between regions of the United States. The modernization of the grid to accommodate today's power flows, serve reliability needs, and meet future projected uses is leading to the incorporation of information processing capabilities for power system controls and operations monitoring. The "Smart Grid" is the name given to the evolving electric power network as new information technology systems and capabilities are incorporated. While these new components may add to the ability to control power flows and enhance the efficiency of grid operations, they also potentially increase the susceptibility of the grid to cyber (i.e., computer-related) attack since they are built around microprocessor devices whose basic functions are controlled by software programming and subject to manipulation over a network.

The information processing attributes which make the smarter grid attractive are the very same attributes which can increase the vulnerability of the electric power system and its critical infrastructure to a potentially devastating cyber attack. The potential of the Smart Grid to revolutionize the ways power is generated and used is great, but so too are the potential risks if not sufficiently addressed.

The Energy Independence and Security Act of 2007[9] (EISA) (P.L. 110-140) defined the attributes of a Smart Grid and plans for its development. The potential for a major disruption or widespread damage to the nation's power system from a large scale cyberattack has increased focus on the cybersecurity of the Smart Grid. While a nation-state might have the resources and capability to create widescale disruption of the U.S. grid (to create economic disorder or as part of a physical attack), terrorists or hackers are less likely to have the sophistication or resources necessary to inflict damage on a national scale. But the possibility of collaboration between terrorists and hackers could increase the capability of these groups to launch network-based cyberattacks against critical U.S. infrastructure.[10]

Among the issues for Congress is whether the existing requirements of current law adequately assure the cybersecurity and reliable operation of the U.S. electricity grid today (and the Smart Grid of tomorrow). This report will look at issues key to these determinations, discussing the scope and point of application of regulations, and where the overall responsibility rests for the cybersecurity of the grid.

ATTRIBUTES OF A SMART GRID

The U.S. Department of Energy (DOE) summarized its view of the potential of the Smart Grid by the year 2030 as:

> ... a fully automated power delivery network that monitors and controls every customer and node, ensuring a two-way flow of electricity and information between the power plant and the appliance, and all points in between. Its distributed intelligence, coupled with broadband communications and automated control systems, enables real-time market transactions and seamless interfaces among people, buildings, industrial plants, generation facilities, and the electric network.[11]

A “smarter” grid is gradually being introduced as upgrades to the systems and components of the power generation, transmission, and distribution infrastructure. Substation and distribution substations are being automated with superior switching capabilities to enhance current flows and control of the grid. Devices called “phasor measurement units” are also being added to substations to make time and location-specific measurements of transmission line voltage, current, and frequency (i.e., synchrophasor measurements made 30 times per second instead of data measured once every two to four seconds by current industrial control systems) providing better tools to improve power system reliability.[12] DOE’s strategic plan targets the deployment of 1,000 phasor measurement units by 2013.[13]

Advanced meters are being introduced in some utility markets which can offer customers the ability to leverage demand response[14] regimes and technologies, and potentially benefit from the interactive features of the Smart Grid by shifting their energy use to off-peak hours. Expectations vary of what a Smart Grid could accomplish, and the estimated costs of a system rise with the increased scope and attributes of a system. Some see the Smart Grid as only an enhancement of today’s existing transmission and distribution systems focused on improving the reliability and efficiency of the grid. Others see the Smart Grid of the future as an integrated system spanning the nation from coast to coast, able to seamlessly combine distributed resources and central power stations across the three major interconnections[15] of the U.S. grid. Under such a vision, distributed and renewable energy resources could be efficiently integrated into the grid, with power from (for example) intermittent wind generation channeled by sensors and intelligent electronics from multiple widely dispersed sites to where power is needed anywhere on the grid. *Figure*

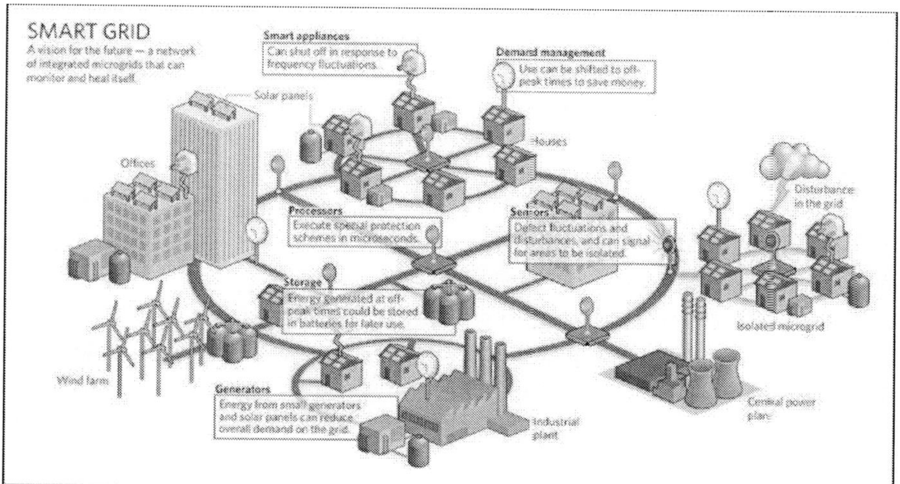

Source: Consumer Energy Report. See http://www.consumerenergyreport.com/wp-content/uploads/2010/04/ smartgrid.jpg.

Figure 2. One Conceptualization of the Smart Grid.

2 shows a possible configuration of a Smart Grid system illustrating components of an interconnected, two-way intelligent network, and distributed and central station sources of power. The efficiency and economy of all grid operations could conceivably be optimized by similarly harnessing all power generation to take advantage of a wide range of generation and storage resources across the United States.

DEFINING CYBERSECURITY FOR THE SMART GRID

A prolonged disruption of energy from the grid was described in a report for the Presidential Commission on Critical Infrastructure[16] as an event which would seriously affect every infrastructure, with the possibility of a cyberattack causing such damage considered as a possibility.

> The widespread and increasing use of Supervisory Control and Data Acquisition17 (SCADA) systems for control of energy systems provides increasing ability to cause serious damage and disruption by cyber means. The exponential growth of information system networks that interconnect the business, administrative, and operational systems contributes to system vulnerability.[18]

EISA gave the National Institute of Standards and Technology (NIST) the role of coordinating the development of a framework to enable the development of the Smart Grid in a safe and secure manner, and NIST issued its first guidelines for cybersecurity of the Smart Grid in 2010. Because cybersecurity threats were perceived as "diverse and evolving," NIST advocated a defense-indepth strategy with multiple levels of security because no single security measure could counter all types of threats.[19] The key to NIST's suggested approach is the determination of risk (i.e., the potential for an unwanted outcome resulting from internal or external factors, as determined from the likelihood of occurrence and the associated consequences) as quantified by the threat (e.g., event, actor or action with potential to do harm), the vulnerability (e.g., weakness in the system), and the consequences (e.g., physical impacts) to the system.[20]

Cybersecurity Vulnerabilities of the Smart Grid

A smarter grid is largely the result of the modernization of the grid. Most sectors of business and industry have gone from the analog age of knobs and dials to the digital age of sensors and electronic displays in the recent past, and that process is now reaching into the nation's electric power infrastructure. Part of the delay has been due to the long-lived nature of the capital assets which make up the industry. The power plants and other expensive components of the grid can function for many productive years if maintenance of the systems is kept up. However, much of the delay in modernization has been due to cost concerns, as many electric utilities seek to manage expenses by delaying replacement of aging systems as long as possible. Several of the better-known vulnerabilities of a smarter grid are discussed in the following paragraphs.

Control Systems

Industrial control (IC) systems are particularly vulnerable to cyberattack because of their intelligence and communications capabilities. IC systems perform a number of functions in the electrical grid, ranging from microprocessor-based control systems which control the actuation and operation of one or more devices, to more sophisticated industrial IC systems which can manage entire industrial processes or automated systems. SCADA systems are a well-known application of remote IC used to monitor and control electric transmission system components.

While cyber-intrusions into the U.S. grid have been reported in recent years,[21] no impairment or other damage has been publicly reported as a result of the attacks. By exploiting loopholes in cybersecurity, cyberattackers could breach the privacy of customers power usage data and access large numbers of meters, perhaps sending deliberately misleading information to the grid. This could potentially overload systems or cause grid operators to respond to false readings. However, concerns exist as to the potential damage that could result in the future from malware left behind by such intrusions or doorways created in systems which could be exploited.[22] The revelation of the complexities of the Stuxnet worm[23] and the alleged targeting of the control systems of a nuclear power plant in Iran[24] have raised additional concerns about the vulnerability of electric power systems worldwide.

The Stuxnet worm reportedly affected the logical decision making functions of the control systems of one particular type of manufacturing facility. Theoretically, Stuxnet-like malware could be adapted to impair other types of process or control systems in malicious or unpredictable ways. Malware modified in such ways could overload and damage targeted equipment and systems in critical infrastructure.[25] The Department of Homeland Security (DHS) conducted a simulated cyberattack on an electric generator control room and partially destroyed a large diesel-electric generator secured for the test as a demonstration of existing power system vulnerabilities.[26]

Connections through other Systems

Utilities have routinely accessed control systems either wirelessly or by telephone lines for many years. In fact, telephone lines often used to access corporate systems may even provide access to utility control systems. A recent GAO report highlighted vulnerabilities at the Tennessee Valley Authority, a large federal utility in the United States.

> ... the interconnections between TVA's control system networks and its corporate network increased the risk that security weaknesses, on the corporate network could affect control systems networks and we determined that the control systems were at increased risk of unauthorized modification or disruption by both internal and external threats.[27]

Communications and Internet Access

Grid devices capable of two-way communications are considered to be potential points of unauthorized system access, and can represent a potential cybersecurity vulnerability. While security protocols may exist to prevent unauthorized entry, wireless networks can be monitored and potentially hacked by cybercriminals. Smart meters are another example of new applications in which the security of data has been mentioned as a concern. At the heart of smart meters are semiconductor chips which allow data about energy use to be sent wirelessly to the electric distribution company, and which potentially allows the meter to control the flow of power to customers. The security of the encrypted information in the communications protocol used by many of these devices has been questioned in the past, and questions have been raised about the effectiveness of efforts to patch identified flaws. The primary consideration in the choice of wireless protocol by semiconductor providers seems to be cost of the system, a criteria which is not always compatible with cybersecurity goals for secure communications.[28] While the developing Smart Grid is likely to use today's internet and wireless communications systems for many years to come, security concerns and data usage requirements may move the system to dedicated communications and information channels serving uniquely Smart Grid uses.

Potential access to SCADA systems via the Internet has often been cited as a key vulnerability for electric power systems. Hackers have used the Internet to launch denial of service,[29] phishing,[30] and various other forms of cyberattack. The effective recording, processing, and exchanging of data is becoming increasingly critical to the reliability of the power system,[31] and deliberately introduced misinformation can be damaging. Internet-linked communications systems may be important to today's interconnected grid, but the Internet also provides a ready path to cyberattack from any corner of the world wide web.

Combined Cyber- and Physical Attacks

The vulnerability of key components of the grid to a physical attack has been long known, but a coordinated and sustained physical and/or cyberattack[32] could be problematic. After a cyberattack, restored systems may be repaired and protected against known vulnerabilities, but a coordinated cyberattack over a short time period targeting different Smart Grid systems could slow recovery. The same would hold true for repeated physical attacks, or combined cyber- and physical attacks on key grid components. Under such circumstances, the resources of emergency programs in place to replace important grid systems and components (such as the Edison Electric Institute's "Spare Transformer Equipment Program"[33]) may not be adequate to provide enough replacement parts to fully restore reliable grid operations in a timely manner.

Supply Chain

The components of smarter grid devices present another potential vulnerability concern. Most of the smart meter, sensor, and other equipment makers are international companies who obtain their components from international sources. Taiwan, Singapore, China, and South Korea are among the largest overseas manufacturers of semiconductors and microprocessors for smart devices. The reliable operation of semiconductor and microprocessor-based devices is based on the low-level firmware[34] controlling the device's basic functions. Firmware in the form of fixed machine-language binary code is found in almost all the electronic devices making up smarter grid products such as programmable controllers and programmable logic arrays. If a hacker or cybercriminal gained access to such devices (especially during the manufacturing process), a section of code could be covertly inserted in the device and activated in such a way as to impair its functioning in a reliable manner. Some might suggest random or statistically based testing of the firmware in smarter grid devices. But the impairment would not need to be placed in all such devices coming off an assembly line. If a large enough sample was impaired, the effect might be enough to cast doubt on the reliability of a whole class of such devices.

Geomagnetic Impairment

An electromagnetic pulse (EMP) is a large burst of electromagnetic radiation resulting from the detonation of a nuclear weapon or from other means. EMPs can cause voltage and current surges which can severely damage electrical equipment and systems. A high-altitude nuclear explosion could cause an EMP which could severely impair an electric power system not shielded from the effects of an EMP.

However, deliberate, large-scale, man-made attacks are not the only potential threat which could take down a substantial portion of the electric grid. Radiation or charged particles ejected into space weather from solar flare eruptions could cause geomagnetic storms which could interfere with telecommunications networks or damage orbiting satellites. Periodically, these eruptions from the sun are powerful enough to disrupt the operations of electric power systems. Solar flares occur in cycles of almost 11 years. [35] Occasionally, severe space weather results in a huge eruption of particles which can be particularly disruptive to electric power systems, and induce ground currents powerful enough to actually melt the copper windings of electric power transformers. The last major solar flare eruption in 1989 caused blackouts across the Canadian province of Quebec for nine hours. But even greater storms can occur perhaps every 100 or more years. The first major eruption of this size was observed in 1859. It destroyed much of the world's telegraph infrastructure of the time. Another event observed in 1921 demonstrates that although these events are rare, they are likely to occur again and could have tremendous societal and economic impacts for the United States:

> ... the occurrence today of an event like the 1921 storm would result in large-scale blackouts affecting more than 130 million people and would expose more than 350 transformers to the risk of permanent damage.[36]

It could take weeks or months to replace most of the damaged transformers. Some of the larger units could take years to replace due to the fact there are no U.S. power transformer manufacturers,[37] and a multi-year backlog exists for the larger units.

CURRENT CYBERSECURITY POLICY FOR THE SMART GRID

A smarter grid is increasing reliance on automated systems to manage energy sources, power control devices, system information, and power flows. Automation will allow information from smarter processors and sensors to be processed at much faster rates, giving power system operators more accurate and timely information, and giving customers greater input into their own power use. But, the possibility that such systems could be highjacked in such a way as to bring down the grid regionally or even nationally, and cause long-term damage to critical systems and components, is giving Congress reason to consider the cybersecurity of the electric power infrastructure.

The passage of the Homeland Security Act of 2002 (P.L. 107-296) (HSA) established the Department of Homeland Security. HSA increased penalties for certain computer-based crimes, and in addition to improving on the earlier Computer Fraud and Abuse Act of 1986,[38] made critical infrastructure protection a key mission of DHS.

Federal efforts to enhance the cybersecurity of the electrical grid were reemphasized with the recognition of cybersecurity as a critical issue for electric utilities in developing the Smart Grid. The Energy Independence and Security Act (EISA) of 2007 (P.L. 110-140) thus added requirements for "a reliable and secure electricity infrastructure" with regard to Smart Grid development. EISA directed NIST to develop a framework for protocols and standards for the Smart Grid to achieve "interoperability"[39] of devices and systems. The American Recovery and Reinvestment Act of 2009 (P.L. 111-5) provided NIST with funds to proceed with this mandate.

Status of Regulatory Implementation

NIST issued a report, "NIST Framework and Roadmap for Smart Grid Interoperability Standards" (NIST SP 1108, January 2010), listing 75 existing standards which are likely to be applicable for an interoperable Smart Grid, and identified 15 high-priority gaps and harmonization issues. NIST also issued a "Smart Grid Cybersecurity Strategy and Requirements" report (NISTIR 7628, September 2010) assessing requirements to ensure the security and reliability of a modernized grid. This report also classified Smart Grid interfaces according to potential impacts that could result from compromise of these systems.

The Federal Energy Regulatory Commission (FERC or the Commission) received primary responsibility for the reliability of the bulk power[40] system from the Energy Policy Act of 2005 (EPACT05) (P.L. 109-58). EPACT05 also gave FERC the authority to commission an "Electric Reliability Organization" (ERO) to establish and enforce reliability standards. FERC subsequently designated the North American Electric Reliability Corporation (NERC) as the ERO. Compliance with reliability standards for electric utilities thus changed from a voluntary, peer-driven undertaking to a mandatory function. While NERC is responsible for standards for the bulk power system in the United

States, it is also seeking recognition by applicable governmental authorities in Canada and Mexico to establish and coordinate reliability standards for the bulk power systems of these respective countries.[41] NERC is also responsible for standards for critical infrastructure protection (CIP) which focus on planning and procedures for the physical security of the grid. NERC has delegated much of the daily process of maintaining reliability of the grid to regional entities via delegation agreements which are subject to FERC approval. Compliance and enforcement with NERC standards is therefore largely a responsibility of these regional entities, which are mostly responsible for imposing penalties for violation of reliability standards.[42] NERC's CIP standards essentially approach cybersecurity as an extension of reliability, and address nine areas including *Sabotage Reporting*, *Critical Cyber Asset Identification*, and *Electronic Security Perimeters*, and *Recovery Plans for Critical Cyber Assets*.[43] These standards identify what utilities must do to protect "cyber assets"[44] seen as critical to the reliability of the bulk power[45] system.

EISA directed FERC to work with NIST in developing Smart Grid interoperability standards, and required FERC to institute a formal rulemaking process to adopt interoperability (including cybersecurity) standards when FERC was satisfied that a consensus on defining the standards had been reached. NIST advised FERC in October 2010 that it had identified five "foundational" sets of standards[46] which may allow FERC to go forward with instituting rulemaking proceedings to adopt standards to ensure Smart Grid functionality, interoperability, and cybersecurity. The standards would be updated as Smart Grid requirements and technologies evolve.

FERC held a technical conference in January 2011 to determine if there was "sufficient consensus" that the five sets of standards proposed by NIST were adequate for standards and protocols to be adopted. The comments of conference participants showed that a consensus on the standards had not in fact been reached with several participants stating that they were unsure of what would constitute a consensus, and questioning whether such standards would then automatically become "mandatory and enforceable." FERC issued a supplemental notice in February 2011 (requesting all comments by April 2011) to aid the Commission's determination as to whether a consensus exists on the proposed NIST interoperability standards. Additional questions were asked in the notice to solicit opinions on related topics including the NIST process for determining the readiness of a standard with regards to interoperability and cybersecurity, whether making standards enforceable would achieve interoperability goals, how testing and certification of cybersecurity requirements could be incorporated into the standards process, and whether the Commission should consider different criteria for evaluating interoperability and functionality (and the extent to which cybersecurity is an element of each).[47]

Assessment of Cybersecurity Progress by Government Accountability Office

Congress asked the U.S. Government Accountability Office (GAO) to assess progress by NIST and FERC on Smart Grid cybersecurity guidelines and standards. GAO issued a report with its findings in January 2011.[48] GAO found that NIST has largely addressed "key cybersecurity elements" such as the cybersecurity risks of Smart Grid systems and had identified security controls essential to such systems. GAO also found that NIST had not addressed the risks of attacks on Smart Grid systems using both cyber- and physical means. GAO recommended that NIST finalize its plan and schedule for updating its cybersecurity guidelines to include these elements.

GAO also pointed out that while EISA had given FERC authority to adopt Smart Grid standards, it did not give FERC specific enforcement authority over implementation of standards. GAO recognized that a regulatory divide exists between federal, state, and local entities on various aspects of Smart Grid interoperability and cybersecurity. As such, GAO states that such standards will remain voluntary unless regulators use other authorities to enforce standards compliance. GAO recommended that FERC develop a coordinated approach (with other regulatory jurisdictions) to monitor voluntary standards and address any gaps in compliance.

FERC and NIST agreed with GAO's recommendations.[49] However, FERC Chairman Wellinghoff stated that while a coordinated approach would be pursued, FERC may lack the authority to address any gaps in compliance, and that this being the case, Congress may want to consider the matter further. Furthermore, in addressing GAO's comments, the Chairman broached the question of whether other-than-voluntary compliance with Smart Grid standards by "relevant manufacturers and utilities" is what Congress intended in EISA.[50]

Meanwhile, NIST has communicated to FERC its own concerns that under the existing process, it could take years for the Commission to adopt rules on the many individual standards needed to achieve interoperability of the Smart Grid while recognizing that many "customized" upgrades to the grid continue to be installed. NIST suggested that in the absence of formal standards, FERC could request "interoperability roadmaps" from utilities to determine if they addressed interoperability based on the NIST framework in the utilities' grid modernization efforts.[51]

Assessment by DOE's Inspector General of FERC's Monitoring of Grid Cybersecurity

The DOE Inspector General released an audit report[52] in January 2011 of FERC's monitoring of grid cybersecurity, with regard to the Commission's responsibilities under EPACT05. The report found that CIP cybersecurity standards developed "did not include a number of security controls commonly recommended for government and industry systems," and criticized FERC's oversight of the process for developing these standards, citing a need for FERC to use its existing authority to ensure timely standards development to address emerging security threats. The report also said that FERC's implementation approach and schedule for CIP standards did not adequately consider risks to information systems, since FERC was focusing on documentation of controls rather than implementation of controls. FERC was advised to revise its focus to ensure that controls to address higher threat risks are given priority. The report noted that these problems existed, in part, because the Commission had "only limited authority" to ensure cybersecurity over the bulk electric system, and could not implement its own reliability standards or issue alerts. However, the report went on to conclude that even "when such authority did exist," FERC did not always act "to ensure that cybersecurity standards were adequate."

FERC Chairman Wellinghoff mostly concurred with the report's recommendations[53] while noting that "the current statutory framework is adequate for addressing emerging cybersecurity threats through mandatory standards." The Chairman stated that FERC will make every effort "to approve CIP standards as soon as possible after they are developed by NERC."[54]

Policy Concerns

Self-determination is a key part of the CIP reliability process. Utilities are allowed to self-identify what they see as "critical assets"[55] under NERC regulations. While compliance with CIP standards became mandatory in 2010, FERC mandated modifications to the criteria defining critical assets for reliability purposes. Only "critical cyber-assets" (i.e., as essential to the reliable operation of critical assets) are subject to CIP standards. FERC directed NERC to revise the standards so that some oversight of the identification process for critical cyber-assets was provided, but any revision is again subject to stakeholder approval. Under current law, FERC cannot issue its own reliability standards.

While reliability standards are mandatory, the ERO process for developing regulations is somewhat unusual in that the regulations are essentially being established by the entities who are being regulated.[56] This can potentially be an issue when cost of compliance is a concern, and acceptable standards may conceivably result from the option with the lowest costs. While FERC ultimately has approval authority over the regulations NERC submits and can remand such regulations it judges as not satisfying requirements, any such revisions are ultimately subject to NERC stakeholder approval. NERC standards could therefore be seen as a minimum threshold for compliance, as some utilities may choose to go beyond what is minimally required. But since utility systems are interconnected in many ways, the system with the least protected network potentially provides the weakest point of access.

OTHER ISSUES IN FEDERAL SMART GRID CYBERSECURITY POLICY

Legacy Systems

Legacy devices and systems (such as many SCADA and substation automation systems) may represent as much of a vulnerability to cybersecurity as new Smart Grid components. They were not designed with cybersecurity in mind, and are often interconnected either via the Internet or by other, sometimes "unsecured" avenues. DOE describes the legacy system communications issue as follows:

> The legacy communication methods that now support grid operations also provide potential cyber attack paths. Many cyber vulnerabilities have been identified by cyber security assessments of SCADA systems. Power grid substation automation and security have also recently been evaluated. The level of automation in substations is increasing, which can lead to more cyber security issues. The cyber security issues identified during the assessments and evaluations need to be resolved, but the known issues should not be construed as a complete assessment of the current power grid security posture.[57]

Some legacy grid devices are being retrofitted with communications capabilities to allow them functionality in the smarter grid, or permit easier maintenance, potentially introducing grid vulnerabilities which may not have existed before. NIST has suggested the establishment of "Priority Actions Plans" to resolve interoperability issues between legacy equipment and Smart Grid systems as these are installed.[58]

FERC specifically rejected proposals by commenters for additional interoperability standards (at this time) addressing existing resources or equipment and cost-effective integration of legacy equipment, and interfaces between utilities.[59] Replacement of these systems is recognized by FERC as a possible option for utilities to apply for recovery of the "stranded costs" of legacy systems replaced by Smart Grid investments. [60]

Leadership of Federal Authority for Smart Grid Cybersecurity

The issue of which, or whether, a single federal agency should have ultimate authority over Smart Grid cybersecurity remains largely unresolved. DHS, DOE, and FERC have all either stated claims or been mentioned as candidates for overall leadership of the issue based upon implications to national security, energy policy and technology leadership, or industry understanding and organizational mission, but the current missions of these agencies is only the starting point for a discussion. Since Smart Grid devices are being installed beyond the federally regulated power grid, the system could potentially face intrusions from cyberattacks initiated on less-protected distribution systems. Leadership of Smart Grid and cybersecurity issues may require the ability to communicate issues and push solutions in forums beyond federal jurisdictions if the goal is to protect against cyber-threats to the grid for the entire United States.

DHS's mission to protect national security against cyber-threats currently centers on "civilian government systems" and DHS works with industry to "secure critical infrastructure and information systems." DHS is focused on analyzing cyber-threats and reducing vulnerabilities, and distributing information on potential cyberattacks to ensure the safety of cybersystems.[61] DHS will distribute warnings of cyber-threats to other federal agencies, state and local governments and to industry.

DOE has the technical and research capabilities of the national laboratories to draw upon with regard to cybersecurity and the Smart Grid, as well as its own internal policy offices which deal with an array of electricity system issues. DOE has worked with electric utilities to understand the cyber-vulnerabilities of the grid, and has tasked the national laboratories with helping to develop technologies and software patches to identify problems and harden legacy control systems against potential cyber-intrusions.

FERC is an independent, self-funded regulatory agency within DOE, with statutory authority over major aspects of the wholesale electric, natural gas, hydroelectric, and oil pipeline industries.[62] As such, FERC's responsibility for the reliability of the grid and related ratemaking authority allow the Commission to discover issues and authorize utilities to recover the jurisdictional portion of costs related to grid critical infrastructure protection upgrades. FERC recognizes that EISA does not make Smart Grid standards mandatory, nor does EISA give it authority to make or enforce such standards.[63] FERC has jurisdiction over the bulk power system, but this currently excludes authority over the nation's electric distribution systems where much of the smarter grid will be deployed. The Commission appears to view its reliability mandate under EISA as giving it the authority to adopt Smart Grid standards which will be applicable to all electric power facilities and devices with Smart Grid features, including those at the distribution level. FERC further ties cybersecurity to reliability by proposing a "harmonization" of each Smart Grid protocol and reliability standard developed.[64]

Cost Challenges of Maintaining a Cyber-Secure Smart Grid

Cybersecurity threats present a constantly moving and growing need for mitigation activities. As new devices and systems are being developed for the Smart Grid, so too is the array of infrastructure elements which must potentially be protected. This is a key concern for utilities, as the resources which may have to be applied to cybersecurity concerns could likewise spiral upward in growth and costs. Utilities may understand how a smarter grid can enhance their efficiency and reliability, and the need for cybersecurity safeguards. Commercial and industrial customers may also see potential benefits in Smart Grid-enabled programs (such as demand response), and it is likely they will fund a significant portion of smarter grid improvements through rates. But at present very few consumers at the residential level seem to understand the potential of the Smart Grid, or have reason to believe it can do anything for them except increase rates.[65] One recent study expects cybersecurity spending to represent approximately 15% of total Smart Grid capital investment between 2010 and 2015 (with total investment estimated to be $1.5 billion in North America).[66]

Smart Grid enhancements across the United States may vary by utility, state, region, and regulatory jurisdiction. Recovery of costs may present a major challenge especially to distribution utilities and state commissions charged with overseeing utility costs. When faced with such decisions, the perception of the potential risks to the utility's local system will likely bear on what costs the utility's ratepayers should pay. EISA requires states only to consider recovery of costs related to Smart Grid systems. FERC has jurisdiction over the bulk power grid, i.e., transmission and wholesale rates in interstate commerce. FERC cannot compel entities involved in distribution to comply with its regulations. Since distribution utilities are regulated by state or local public utility commissions, Smart Grid investments will likely require approval by these jurisdictions. The issue of whether Smart Grid standards for interoperability should be voluntary or mandatory for these jurisdictions remains unresolved.

Privacy and Data Security

The sharing of information in applications used by the smarter grid has raised questions on the safety of that information. Security of customer information in wireless applications, and how personal data characteristics (such as customer usage information) can be protected are issues often mentioned in discussions of the Smart Grid and cybersecurity. Encryption of data (with limited decryption for data checking), and aggregation of data at high levels to mask individual usership have been mentioned as ways to protect the identity of individual customers. Limiting the amount of data to just information needed for billing purposes has been suggested as a remedy to the potential for misuse of data.[67] Real-time monitoring of these networks may alert system operators to suspicious activities. But even these methods might not be enough to guard against a sophisticated intruder if data security is taken for granted. The development of Smart Grid standards for electric utilities and their partners governing the collection and use of customer data may be a possible next step.

If customers participate in demand-side management programs, then customer usage data can provide a wealth of information for a variety of programs for interruptible loads or time-of-use rates. But customer-specific data stored in home area networks (HANs), or customer-specific data communicated between the HAN and distribution utility (or load aggregator) must be secure to protect the privacy of information. The advent of electric vehicles may offer another potential payload of data on customer movement and habits, if data collected or stored is not restricted to electricity consumption for billing purposes.

The National Regulatory Research Institute recommended in a recent report[68] that public utility commissions should define the information which utilities will collect, determining with whom and for what purpose it should be shared, and assess the need for protecting the data. While allowing states to develop their own policies may be accomplished more quickly, the report also advocates for a national approach to Smart Grid privacy issues.

DOE has worked with electric utilities and the Electric Power Research Institute (EPRI) to develop profiles for Smart Grid products to address concerns about data privacy. These profiles are meant to provide guidance on how to build cybersecurity into Smart Grid applications under development, so that they can be more securely integrated into the Smart Grid.[69]

Government-Industry Coordination on Threats

In 2010, the Edison Electric Institute (EEI) issued its own "Principles for Cyber Security and Critical Infrastructure Protection"[70] to discuss development of a framework for addressing cybersecurity threats. EEI, the trade association of U.S. shareholder-owned electric utilities, suggested a coordinated effort of government, industry, and suppliers of critical grid systems and components was necessary to protect the grid from cyberattacks as a "shared responsibility." EEI believes that the sharing of intelligence on evolving threats and a clear plan of action on imminent threats is key to this arrangement. EEI also expects that such a policy would help protect national security and the public welfare, and aid in the prioritization of assets which need enhanced security. EEI expects that such a partnership could "utilize all stakeholder's expertise," allowing owners and operators of critical infrastructure to propose mitigation strategies that will avoid significant adverse consequences to utility operation or assets.

OTHER COLLABORATIVE CYBERSECURITY INITIATIVES

There are a number of collaborative efforts involving private sector, academia, national laboratories, consumer groups, and others looking at cybersecurity and Smart Grid issues. Federal agencies are also working together to ensure communication and share resources on these issues, and helping to fund industry cybersecurity initiatives.[71] Two of the major DOE-led initiatives are discussed below.

DOE is home to the Federal Smart Grid Task Force (FSGTF) established under EISA to "ensure awareness, coordination, and integration of the diverse activities of DOE's Office of Electricity Delivery and Energy Reliability (OE) and elsewhere in the federal government related to Smart Grid technologies, practices, and services."[72] FERC, the Department of Commerce, the Environmental Protection Agency, the Department of Agriculture, and the Department of Defense are also members of the FSGTF.

DOE also funds (with OE leading DOE's participation) the "Trustworthy Cyber Infrastructure for the Power Grid" (TCIPG) initiative, working with DHS and industry.[73] TCIPG is led by the academic sector as a public-private research partnership, supporting the development of resilient and secure Smart Grid systems, leveraging research previously funded by the National Science Foundation:

> TCIPG's research plan is focused on securing the low-level devices, communications, and data systems that make up the power grid, to ensure trustworthy operation during normal conditions, cyber-attacks, and/or power emergencies. At the device level, new key functionality is being designed in hardware in order to detect attacks and failures and to restore proper system operation. Likewise, virtual machine technology is being developed and adapted for advanced power meters in order to permit new power use scenarios while preserving privacy. At the protocol level, new techniques are being developed to detect, react to, and recover from cyber attacks that occur while preserving integrity, availability, and real-time requirements.

Participating in TCIPG are the University of Illinois at Urbana-Champaign, Dartmouth College, Cornell University, the University of California at Davis, and Washington State University. Current TCIPG projects are focused on developing "trustworthy cyber infrastructure and technologies" for active demand management and wide-area monitoring and control, response to and managing cyber events, and risk/security assessments.

CURRENT LEGISLATION

Several bills introduced in the 112th Congress addressing the cybersecurity of the Smart Grid are summarized below.

Senate

The Cyber Security and American Cyber Competitiveness Act of 2011[74] (S. 21) would secure the United States against cyberattack and protect sensitive information and the identities of U.S. citizens and business. The bill focuses on the increasing dependence of the United States on information technology, and highlights a growing need for public awareness and preparedness of cyber-threats.

The Cybersecurity and Internet Freedom Act of 2011[75] (S. 413) is intended to enhance the "security and resiliency of the cyber and communications infrastructure" of the United States. Among its provisions, the bill would establish within DHS a National Center for Cybersecurity and Communications (NCCC) to protect the federal and national information infrastructure. The duties of the NCCC Director would include the identification and evaluation of cyber-risks to critical infrastructure (covered by the bill) on a continuous, sector-by-sector basis. The NCCC would also issue regulations establishing risk-based security performance requirements to secure covered critical infrastructure against cyber risks. Each owner or operator of covered critical infrastructure would be required to certify to the NCCC Director that approved security and emergency measures had been implemented. Civil penalties for non-compliance would be established by the bill. The President would be authorized by the bill to declare a national cyber-emergency, and would be required to notify owners and operators of critical infrastructure about the nature of a cyber-threat if a national cyber-emergency was declared.

The Cyber Security Public Awareness Act of 2011[76] (S. 813) is intended to promote the awareness of cybersecurity. Among its provisions, the bill requires a report from be DHS on technical options to enhance the security of information networks of entities which own or manage critical infrastructure. The bill also requires a report from the Secretary of Homeland Security in consultation with the Secretary of Defense and the National Intelligence Director concerning the threat posed by a cyberattack to disrupt the national electrical grid and its implications to national security, the options for quick recovery to provide for national security as well as for restoration of all electrical service in the United States, and the development of a plan to prevent disruption of the U.S. grid by a cyberattack.

House

The National Infrastructure Development Bank Act of 2011[77] (H.R. 402) aims to facilitate infrastructure development. The bill identifies the nation's energy infrastructure as an area in need of modernization. Eligibility is extended to Smart Grid projects as energy infrastructure development.

The Secure High-voltage Infrastructure for Electricity from Lethal Damage Act[78] (H.R. 668) seeks to protect the bulk power system and electric infrastructure critical to the defense and wellbeing of the United States from EMP threats (such as nuclear weapons), or natural geomagnetic disturbance (GMD) vulnerabilities (such as solar flares). The bill proposes to authorize FERC to order the ERO to take emergency measures to protect the reliability of the bulk power system upon identification of an imminent threat to grid security from an EMP attack or GMD event. FERC would be authorized to allow utilities to recover "substantial costs" incurred from compliance with such an emergency order. The bill would also authorize FERC to issue rules or orders to protect against such grid security vulnerabilities not covered by ERO reliability standards (and would rescind the FERC rule or order when a sufficient ERO reliability standard was in place). FERC would be required to direct the ERO to propose reliability standards to protect against "reasonably foreseeable" EMP or GMD events. The ERO would also be required to propose reliability standards (balancing risks and costs) to address availability of large transformers capable of restoring reliable grid operations after an EMP or GMD event, providing entities (either individually or jointly) which own such transformers to ensure their adequate availability in case such transformers are destroyed or disabled by such an EMP or GMD event. The President would be required to identify up to 100 electric infrastructure facilities critical in supplying power for the defense of the United States which would be subject to FERC rules and orders related to EMP or GMD events. Recovery of full incremental costs incurred in compliance with such FERC rules and orders would be allowed for owners of the identified defense critical electric facilities.

White House Legislative Proposal on Cyberspace

The Obama administration has launched its own comprehensive proposals for legislation on cybersecurity.[79] Protection of critical infrastructure (including the electric grid, financial systems, and transportation networks) is among the main proposals. Other initiatives are suggested to better report cyber-breaches so that consumers are aware that their personal data may have been accessed by intruders, and "synchronized" penalties for cyber-crime with mandatory minimum punishments for cyber-intrusions into critical infrastructure.

The Administration wants better preparedness and coordination between federal agencies, state governments, and industry in order to prevent cyberattacks against critical infrastructure. The proposals include development of cybersecurity standards and enhanced information sharing about cyber threats. The Administration also proposes legislative provisions for DHS to help industry, states, and local government to recover from a cyber-intrusion and fix the resulting damage, or provide advice on building a better network. DHS would also review cyber-protection strategies and requirements, making public its findings on readiness (while not revealing the specific nature of any deficiencies).

ADDITIONAL COMMENTS

Much of the federal efforts related to the cybersecurity of the smarter grid relate to efforts to stay ahead of cyber-threats. But such is the complexity of the cyber-threat universe (with defined threats, perceived weaknesses, and the possibility of continual upgrades of both hardware and software systems) that staying ahead of the threats could quickly evolve into a rapidly growing, never-ending cost center for utilities. While a key to cost control is in assessment of the likely risks, cost recovery for reliability under current federal law is based on mandatory compliance with CIP cybersecurity standards for reliability purposes.

Cost recovery for Smart Grid expenditures or upgrades is voluntary under current law. Such discretionary treatment may possibly result in liability claims if damages result from an attack on Smart Grid systems. While the "Support Anti-Terrorism by Fostering Effective Technologies Act" of 2002[80] provides "risk management" and "litigation management" protection for qualified antiterrorism technologies and others in the supply and distribution chain, such protections likely do not apply to electric utilities as adopters of Smart Grid technologies and devices.

Current legislative and Administration proposals largely speak to a prioritization of critical infrastructure facilities, with consideration to how these relate to national security, and national economic recovery. However, recoverability from a cyberattack on the scale of something which might take down a significant portion of the grid will likely be very difficult. Maintaining a ready inventory of critical spare parts in close proximity to key installations would likely prove useful to quickening recovery efforts from some types of attack.

The current Smart Grid cybersecurity discussions largely focuses on the security of central station power plants and transmission systems. However, the future Smart Grid may increasingly depend on renewable energy, fuel cells, and other distributed resources like energy storage as these technologies are increasingly integrated into the nation's energy framework. The development of the Smart Grid with distributed and renewable power generation resources may add a level of security to the grid, since these resources do not have the fuel requirements of fossil generation. Damage to the fossil fuel delivery networks would likely impair operation of central station generating plants, depending on how much of an inventory of fuel is stored on-site. This greater diversity of resource options would likely further enhance the Smart Grid's expected improvement in reliability due a greater diversity of resource options, joining together these newer elements with traditional power stations in the power grid of the future. But the characteristics that these elements bring to the system could be considered in the design of CIP standards and protocols.

Although this report has focused on cybersecurity aspects of a smarter electrical grid, technologies being developed for use by the Smart Grid could also be used by natural gas pipeline, water supply, and telecommunications systems. Consideration could be given to applying similar control system device and system safeguards to these other vital utility systems. Consideration could be given to applying similar control system device and system safeguards to these other vital utility systems. The electricity grid is connected to (and largely dependent on) these other resource and communications providers as the vast majority of power production today needs water for steam generation or natural gas for fuel. Similarly, natural gas pipelines use many electrically powered compressors to maintain pipeline pressure and move gas through the system, and the nation's water systems is a notable consumer of electricity for distribution and supply purposes. These interdependent systems are also vulnerable to natural hazards such as earthquakes and tornadoes. The remote sensing and other technologies expected to be employed by the Smart Grid could also help mitigate the effects of regional natural disasters on critical infrastructure, as these are expected to be capable of detecting outages and optimizing power flows to minimize damage and enable a more complete recovery.

End Notes

[1] EIA, Glossary, http://www.eia.gov/tools/glossary/index.cfm?id=B.

[2] EIA, Glossary, http://www.eia.gov/tools/glossary/index.cfm?id=B.

[3] As of 2007, there were 210 investor-owned electric utilities, 2,009 publicly-owned electric utilities, 883 consumerowned rural electric cooperatives, and nine federal electric utilities. Energy Information Administration (EIA), Electric Power Industry Overview 2007, http://www.eia.doe.gov/electricity/page/prim2/toc2.html.

[4] The bulk power system makes it possible for utilities to engage in wholesale (sales for resale) electric power trade. U.S. Energy Information Administration, Overview—Power Transactions & Interconnected Networks, 2011, http://www.eia.gov/cneaf/electricity/page/prim2/chapter7.html.

[5] Massachusetts Institute of Technology, Retrofitting of Coal-Fired Power Plants for CO2 Emissions Reductions, March 23, 2009, http://web.mit.edu/mitei/docs/reports/meeting-report.pdf.

[6] Thomas A. Prevost and David J. Woodcock, Transformer Fleet Health and Risk Assessment, Weidman Electrical Technology, IEEE PES Transformers Committee Tutorial, March 13, 2007, http://grouper.ieee.org/groups/transformers/info/S07/S07-TR_LifeExtension.pdf.

[7] K. Anderson, D. Furey, and K. Omar, Frayed Wires: U.S. Transmission System Shows its Age, Fitch Ratings, October 25, 2006.

[8] Key legislation include the Public Utility Regulatory Policies Act of 1978, the Energy Policy Acts of 1992 and 2005, and the Energy Independence and Security Act of 2007.

[9] See Title XIII at 42 U.S.C. 17381 *et seq.*

[10] CRS Report R41674, *Terrorist Use of the Internet: Information Operations in Cyberspace*, by Catherine A. Theohary and John Rollins.

[11] United States Department of Energy, Office of Electric Transmission and Distribution, *"Grid 2030" A National Vision for Electricity's Second 100 Years*, July 2003, p. 27, http://www.oe.energy.gov/DocumentsandMedia/Electric_Vision_Document.pdf.

[12] Pacific Northwest National Laboratory, *North American SynchroPhasor Initiative*, May 2011, http://www.naspi.org/.

[13] U.S. Department of Energy, *Strategic Plan*, May 2011, http://www.energy.gov/news/documents/DOE_StrategicPlan.pdf.

[14] "Demand Response" allows retail customers to participate in electricity markets by giving them the ability to respond to prices as they change over time—either daily or hourly in most instances. See http://www.oe.energy.gov/demand.htm.

[15] The Eastern, Western and Texas interconnections of the U.S. grid. See http://www.eia.doe.gov/cneaf/electricity/chg_stru_update/fig7.html.

[16] Critical Foundations: Protecting America's Infrastructures, The President's Commission on Critical Infrastructure Protection , October 1997, http://www.fas.org/sgp/library/pccip.pdf. (CF)

[17] A system of remote control and telemetry used to monitor and control the transmission system. See http://www.eia.doe.gov/tools/glossary/index.cfm?id=S.

[18] CF, p. 12.

[19] The Smart Grid Interoperability Panel Cyber Security Working Group, Introduction to NISTIR 7628, National Institute of Standards and Technology, Guidelines for Smart Grid Cyber Security, September 2010, http://csrc.nist.gov/publications/nistir/ir7628/introduction-to-nistir-7628.pdf. (NISTIR)

[20] NISTIR, p. 9.

[21] Siobhan Gordon, Electricity Grid in U.S. Penetrated by Spies, The Wall Street Journal, April 8, 2009, http://online.wsj.com/article/SB123914805204099085.html.

[22] Ibid.

[23] "[A worm is a]n independent computer program that reproduces by copying itself from one system to another across a network. Unlike computer viruses, worms do not require human involvement to propagate." See http://www.gao.gov/new.items/d09661t.pdf, p.6.

[24] "A successful attack by a software application such as the Stuxnet worm could result in manipulation of control system code to the point of inoperability or long-term damage. Should such an incident occur, recovery from the damage to the computer systems programmed to monitor and manage a facility and the physical equipment producing goods or services could be significantly delayed. Depending on the severity of the attack, the interconnected nature of the affected critical infrastructure facilities, and government preparation and response plans, entities and individuals relying on these facilities could be without life sustaining or comforting services for a long period of time. The resulting damage to the nation's critical infrastructure could threaten many aspects of life, including the government's ability to safeguard national security interests." CRS Report R41524, The Stuxnet Computer Worm: Harbinger of an Emerging Warfare Capability, by Paul K. Kerr, John Rollins, and Catherine A. Theohary.

[25] "The Stuxnet trojan/virus is the first publicly known 'worm' to target industrial control systems. The threat posed by Stuxnet has been portrayed as beyond anything seen before. Its goal was to sabotage a real-world industrial plant, not disrupt abstract IT systems. It was aimed at industrial control systems with the intention to reprogram PLCs [programmable logic controllers—devices used for control of processes] in a manner that would sabotage the plant, hiding the changes from programmers or users. Stuxnet has highlighted the potential to directly attack industrial control systems used in critical national infrastructure, including energy, water, and transport sectors." Richard Piggin, Control Network Security Lessons from Stuxnet, Consulting-Specifying Engineer, February 3, 2011, http://m.csemag.com/index.php?id=2832&tx_ttnews%5Btt_news%5D=42715&cHash=42da992b88.

[26] Military Photos.net, AURORA Test Validated Fears of Dept. of Homeland Security, October 1, 2007, http://www.militaryphotos.net/forums/showthread.php?121081-AURORA-test-validated-fears-of-Dept.-of-Homeland-Security.

[27] U.S. Government Accountability Office, INFORMATION SECURITY Cyber Threats and Vulnerabilities Place Federal Systems at Risk, GAO-09-661T, May 5, 2009, p. 12, http://www.gao.gov/new.items/d09661t.pdf.

[28] Jeffrey Carr, "Ember Needs a Wake-Up Call from the CIA," Forbes, January 31, 2011, http://blogs.forbes.com/jeffreycarr/2011/01/31/ember-needs-a-wake-up-call-from-the-cia/.

[29] A method of attack that denies system access to legitimate users without actually having to compromise the targeted system. From a single source, the attack overwhelms the target computers with messages and blocks legitimate traffic. It can prevent one system from being able to exchange data with other systems or prevent the system from using the Internet. See http://www.gao.gov/new.items/d09661t.pdf, p.5.

[30] A high-tech scam that frequently uses spam or pop-up messages to deceive people into disclosing sensitive information. Internet scammers use e-mail bait to "phish" for passwords and financial information from the sea of internet users. See http://www.gao.gov/new.items/d09661t.pdf, p. 6.

[31] NISTIR. Page 7.

[32] James St. Pierre, *Smart Grid Interoperability Panel Cyber Security Working Group*, National Institute of Standards and Technology, Presentation at National Electricity Forum, Washington, DC, February 16-17, 2011, p. 10, http://www.nationalelectricityforum.org/pdfs/NationalElectricityForum_StPierre_2011Feb08.pdf.

[33] Edison Electric Institute, *Spare Transformers*, 2011, http://www.eei.org/ourissues/ElectricityTransmission/Pages/SpareTransformers.aspx.

[34] Software that is embedded in a hardware device that allows reading and executing the software, but does not allow modification, e.g., writing or deleting data by an end user. See http://www.its.bldrdoc.gov/fs-1037/dir-015/_2236.htm.

[35] Committee on the Societal and Economic Impacts of Severe Space Weather Events, National Research Council, *Severe Space Weather Events—Understanding Societal and Economic Impacts: A Workshop Report*, National Academies, May 22-23, 2008, http://books.nap.edu/catalog.php?record_id=12507.

[36] Ibid.

[37] Yousaf M. Butt, "The EMP Threat: Fact, Fiction, and Response," *Space Review*, February 1, 2010, http://www.thespacereview.com/article/1553/2.

[38] 18 U.S.C. § 1030.

[39] Interoperability can be defined as the capability of two or more networks, systems, devices, applications, or components to share and readily use information securely and effectively with little or no inconvenience to the user. GridWise Architecture Council, *Interoperability Path Forward Whitepaper*, November 30, 2005, http://www.gridwiseac.org/pdfs/interoperability_path_whitepaper_v1_0.pdf.

[40] The electrical generation resources, transmission lines, interconnections with neighboring systems, and associated equipment, generally operated at voltages of 100 kV or higher. http://www.nerc.com/files/Glossary_2009April20.pdf.

[41] North American Electric Reliability Corporation, Rules of Procedure, March 21, 2008, http://www.nerc.com/files/NERC_Rules_of_Procedure_EFFECTIVE_20080321.pdf.

[42] The maximum civil penalty which can be assessed for violation of reliability standards is $1 million per day. Revised Policy Statement on Penalty Guidelines, 132 FERC ¶ 61,216, issued September 17, 2010.

[43] North American Electric Reliability Corporation, Reliability Standards for the Bulk Electric Systems of North America, April 20, 2010, http://www.nerc.com/files/Reliability_Standards_Complete_Set.pdf.

[44] Programmable electronic devices and communication networks including hardware, software, and data. http://www.nerc.com/files/Glossary_2009April20.pdf.

[45] In FERC Order No. 743, the North American Electric Reliability Corporation was directed to revise its definition of the term "bulk electric system" to ensure that the definition encompasses all facilities necessary for operating an interconnected electric transmission network. The revision will be made in accordance with NERC's Reliability Standards development process. See http://www.ferc.gov/whats-new/comm-meet/2010/111810/E-2.pdf.

[46] The standards produced by the International Electrotechnical Commission (IEC) focus on the informational models and procedures necessary to promote the reliable and efficient operation of the Smart Grid. Standard IEC 62351 specifically addresses the cyber security of communications protocols.

[47] Federal Energy Regulatory Commission, Smart Grid Interoperability Standards—Supplemental Notice Requesting Comments, February 16, 2011, http://www.ferc.gov/EventCalendar/Files/20110228084004-supplemental-notice.pdf.

[48] U.S. Government Accountability Office, Electricity Grid Modernization: Progress Being Made on Cybersecurity Guidelines, but Key Challenges Remain to be Addressed, GAO-11-117, January 12, 2011, http://www.gao.gov/products/GAO-11-117.

[49] Ibid, p. 27.

[50] Federal Energy Regulatory Commission, Letter from FERC Chairman Wellinghoff to David A. Powner, Director, Information Technology Management Issues for the United States Government Accountability Office, December 23, 2010, http://www.ferc.gov/industries/electric/indus-act/smart-grid/wellinghoff-response-GAO.pdf.

[51] Letter from G. Arnold, NIST to FERC Chairman J. Wellinghoff, dated April 7, 2011. See http://www.nist.gov/smartgrid/upload/FERC_Letter1.pdf.

[52] U.S. Department of Energy, Office of Inspector General, Audit Report, Federal Energy Regulatory Commission's Monitoring of Power Grid Cyber Security, January 2011, http://www.ig.energy.gov/documents/IG-0846.pdf. (DOEIG).

[53] DOEIG, pages 11-12.

[54] DOEIG, pages 21-22.

[55] Defined by NERC as "facilities, systems, and equipment which, if destroyed, degraded, or otherwise rendered unavailable, would affect the reliability or operability of the bulk electric system." See http://www.nerc.com/files/Glossary_of_Terms_2011Mar15.pdf.

[56] NERC works with eight regional entities to improve the reliability of the bulk power system. The members of the regional entities come from all segments of the electric industry: investor-owned utilities; federal power agencies; rural electric cooperatives; state, municipal and provincial utilities; independent power producers; power marketers; and endues customers. These entities account for virtually all the electricity supplied in the United States, Canada, and a portion of Baja California Norte, Mexico. See http://www.nerc.com/page.php?cid=1%7C9%7C119.

[57] U.S. Department of Energy Office of Electricity Delivery and Energy Reliability, Study of Security Attributes of Smart Grid Systems—Current Cyber Security Issues, INL/EXT-09-15500, April 2009, http://www.inl.gov/scada/publications/d/securing_the_smart_grid_current_issues.pdf.

[58] National Institute of Standards and Technology, NIST Framework and Roadmap for Smart Grid Interoperability Standards, Release 1.0, NIST Special Publication 1108, January 2010, http://www.nist.gov/public_affairs/releases/upload/smartgrid_interoperability_final.pdf.

[59] On July 16, 2009, the Commission issued a Smart Grid Policy Statement in which, among other things, the Commission explained its view that EISA does not make any Smart Grid standards mandatory and does not give the Commission authority to make or enforce any

such standards. The Commission clarified that, under current law, its authority, if any, to make Smart Grid standards mandatory [or allow recovery in rates] derives from the Federal Power Act. Smart Grid Policy, 128 FERC ¶ 61,060, issued July 16, 2009. See http://www.ferc.gov/whats-new/comm-meet/2009/071609/E-3.pdf. (FERCSGP)

[60] The Commission also proposed to permit applicants to seek recovery of the otherwise stranded costs of legacy systems that are to be replaced by smart grid equipment. The Commission stated that an appropriate plan for the staged deployment of smart grid equipment, which could include appropriate upgrades to legacy systems where technically feasible and cost-effective, could help minimize the stranding of unamortized costs of legacy systems. The Commission therefore proposed that any request to recover stranded legacy system costs must demonstrate that such a migration plan has been developed. FERCSGP. Page 82.

[61] Department of Homeland Security, Safeguard and Secure Cyberspace, http://www.dhs.gov/xabout/gc_1240609042614.shtm.

[62] Federal Energy Regulatory Commission, FY 2010 Performance and Accountability Report, November 2010, http://www.ferc.gov/about/strat-docs/2010-audit.pdf.

[63] See footnote 58.

[64] FERCSGP, p. 27.

[65] M. D. Garratt, Distributech Highlights Challenge of SmartGrid: What's in it for Consumers?, CleanMakesGreen.com, February 7, 2011, http://cleanmakesgreen.com/2011/02/07/distributech-highlights-challengeof-smartgrid-what%E2%80%99s-in-it-for-consumers/.

[66] Business Wire, Smart Grid Cyber Security Market to Reach $3.7 Billion by 2015, According to Pike Research, Smart Grid Cyber Security, June 23, 2010, http://www.businesswire.com/portal/site/home/permalink/?ndmViewId=news_view&newsId=20100623005613&newsLang=en.

[67] Mathew J. Schwartz, California Proposes Smart Grid Data Privacy Standards, Information Week, May 18, 2011, http://mobile.informationweek.com/10247/show/ 8d96dc971db09af9557e4ad59b4708df&t=f6de813a074007af2879af82d46659ef.

[68] Sherry Lichtenberg, Smart Grid Data: Must There Be Conflict Between Energy Management and Consumer Privacy?, National Regulatory Research Institute, December 2010, http://www.nrri.org/pubs/telecommunications/NRRI_smart_grid_privacy_dec10-17.pdf.

[69] Patricia Hoffman, Statement Before the U.S. Senate Committee on Energy and Natural Resources, Office of Electricity Delivery and Energy Reliability - U.S. Department of Energy, Cybersecurity of the bulk-power system and electric infrastructure, May 5, 2011, p. 5, http://energy.senate.gov/public/_files/HoffmanTestimony050511.pdf.

[70] See http://www.eei.org/ourissues/ElectricityTransmission/Documents/cyber_security_principles.pdf.

[71] Electric Light & Power.com, DOE Picks EPRI Collaborative to Lead Cybersecurity Project, December 27, 2010, http://www.elp.com/index/display/article-display/6433556709/articles/electric-light-power/policy-and_regulation/2010/09/DOE_picks_EPRI_collaborative_to_lead_cybersecurity_project_.html.

[72] U.S. Department of Energy, Federal Smart Grid Task Force, http://www.oe. energy.gov/smartgrid_taskforce.htm.

[73] Trustworthy Cyber Infrastructure for the Power Grid, About TCIPG, 2011, http://tcipg.org/about-tcipg-trustworthycyber-infrastructure-power-grid.

[74] Introduced in January 2011 by Sen. Reid et al.

[75] Introduced in February 2011 by Sen. Lieberman et al.

[76] Introduced in April 2011 by Senators Whitehouse and Kyl.

[77] Introduced in January 2011 by Rep. DeLauro et al.

[78] Introduced in February 2011 by Rep. Franks et al.

[79] White House, Office of the Press Secretary, *Fact Sheet: Cybersecurity Legislative Proposal*, May 12, 2011, http://www.whitehouse.gov/the-press-office/2011/05/12/fact-sheet-cybersecurity -legislative-proposal.

[80] P.L. 107-296.

In: Cybersecurity for the Electric Smart Grid ISBN: 978-1-61942-353-4
Editors: P. Barker and R. F. Price

Chapter 2

Electricity Grid Modernization: Progress Being Made on Cybersecurity Guidelines, but Key Challenges Remain to Be Addressed*

United States Government Accountability Office

Why GAO Did This Study

The electric industry is increasingly incorporating information technology (IT) systems into its operations as part of nationwide efforts—commonly referred to as smart grid— to improve reliability and efficiency. There is concern that if these efforts are not implemented securely, the electric grid could become more vulnerable to attacks and loss of services. To address this concern, the Energy Independence and Security Act of 2007 (EISA) provided the National Institute of Standards and Technology (NIST) and Federal Energy Regulatory Commission (FERC) with responsibilities related to coordinating the development and adoption of smart grid guidelines and standards.

GAO was asked to (1) assess the extent to which NIST has developed smart grid cybersecurity guidelines; (2) evaluate FERC's approach for adopting and monitoring smart grid cybersecurity and other standards; and (3)

* This is an edited, reformatted and augmented version of The United States Government Accountability Office publication, Report to Congressional Requesters GAO-11-117, dated January 2011.

identify challenges associated with smart grid cybersecurity. To do so, GAO analyzed agency documentation, interviewed responsible officials, and hosted an expert panel.

What GAO Recommends

GAO recommends that NIST finalize its plan and schedule for updating its cybersecurity guidelines to incorporate missing elements, and that FERC develop a coordinated approach to monitor voluntary standards and address any gaps in compliance. Both agencies agreed with these recommendations.

What GAO Found

NIST has developed, and issued in August 2010, a first version of its smart grid cybersecurity guidelines. The agency developed the guidelines—for entities such as electric companies involved in implementing smart grid systems—to provide guidance on how to securely implement such systems. In doing this, NIST largely addressed key cybersecurity elements that it had planned to include in the guidelines, such as an assessment of the cybersecurity risks associated with smart grid systems and the identification of security requirements (i.e., controls) essential to securing such systems. This notwithstanding, NIST did not address an important element essential to securing smart grid systems that it had planned to include—addressing the risk of attacks that use both cyber and physical means. NIST also identified other key elements that surfaced during its development of the guidelines that need to be addressed in future guideline updates. NIST officials said that they intend to update the guidelines to address the missing elements, and have drafted a plan to do so. While a positive step, the plan and schedule are still in draft form. Until the missing elements are addressed, there is an increased risk that smart grid implementations will not be secure as otherwise possible.

In 2010, FERC began a process to consider an initial set of smart grid interoperability and cybersecurity standards for adoption, but has not developed a coordinated approach to monitor the extent to which industry is following these standards. While EISA gives FERC authority to adopt smart grid standards, it does not provide FERC with specific enforcement authority. This means that standards will remain voluntary unless regulators are able to use other authorities—such as the ability to oversee the rates electricity providers charge customers—to enforce them. Additionally, although regulatory fragmentation—the divided regulation over aspects of the industry between federal, state, and local entities—complicates oversight of smart grid interoperability and cybersecurity, FERC has not developed an approach coordinated with other regulators to monitor whether industry is following the voluntary smart grid standards it adopts. FERC officials said they have not yet determined whether or how to do so. Nonetheless, adherence to standards is an important step toward achieving an interoperable and secure electricity system and establishing an approach for coordinating on standards adoption could help address gaps, if they arise.

With respect to challenges to securing smart grid systems, GAO identified the following six key challenges:

• Aspects of the regulatory environment may make it difficult to ensure smart grid systems' cybersecurity.	• Consumers are not adequately informed about the benefits, costs, and risks associated with smart grid systems.
• Utilities are focusing on regulatory compliance instead of comprehensive security.	• There is a lack of security features being built into certain smart grid systems.
• The electric industry does not have an effective mechanism for sharing information on cybersecurity.	• The electricity industry does not have metrics for evaluating cybersecurity.

ABBREVIATIONS

DOE	Department of Energy
EISA	Energy Independence and Security Act of 2007
FERC	Federal Energy Regulatory Commission
IT	information technology
NARUC	National Association of Regulatory Utility Commissioners
NERC	North American Electric Reliability Corporation
NIST	National Institute of Standards and Technology
PUC	Public Utility Commission

January 12, 2011

The Honorable Bennie G. Thompson
Ranking Member
Committee on Homeland Security
House of Representatives

The Honorable Yvette D. Clarke
House of Representatives

The electric power industry is increasingly incorporating information technology (IT) systems and networks into its existing infrastructure (e.g., electricity networks including power lines and customer meters) as part of nationwide efforts—commonly referred to as smart grid—aimed at improving reliability and efficiency and facilitating the use of alternative energy sources (e.g., wind, solar). Despite these anticipated benefits, cybersecurity and industry experts have expressed concern that if smart grid systems are not implemented securely, they will be vulnerable to attacks that could result in widespread loss of electrical services essential to maintaining our national economy and security. Experts have also expressed concern about how well smart grid systems will work together (i.e., interoperate), whether modifications will be needed to achieve interoperability, and the extent to which the cost of modifications will be passed to consumers.

The Energy Independence and Security Act of 2007 (EISA)[1] directed the National Institute of Standards and Technology (NIST) to coordinate development of a framework of, among other things, IT standards for ensuring that smart grid systems and networks are interoperable. As part of its efforts to accomplish this, NIST planned to identify interoperability and cybersecurity standards to ensure such systems and networks interoperate properly and are cybersecure. In addition to these undertakings, NIST also identified the need to develop cybersecurity guidelines, for organizations such as electric companies, on how to securely implement smart grid systems. EISA also directed the Federal Energy Regulatory Commission (FERC)—the primary federal regulator of the electricity system—to adopt those standards (identified as part of the NIST efforts) that it deemed necessary to ensure smart grid functionality and interoperability.

As agreed, our objectives were to (1) assess the extent to which NIST has developed smart grid cybersecurity guidelines; (2) evaluate FERC's efforts to adopt smart grid cybersecurity and other standards and monitor their use by industry; and (3) identify challenges associated with ensuring the cybersecurity of the smart grid.

To accomplish the first objective, we analyzed NIST's plans to develop smart grid cybersecurity guidelines; assessed the agency's efforts to date to carry out the plans; and then compared this information to identify any variances, causes, and potential negative impacts; we also interviewed NIST officials responsible for developing the guidelines and industry stakeholders who are to use them. To accomplish the second objective, we collected and analyzed documentation of FERC plans; interviewed FERC officials; and interviewed representatives from seven state electricity regulatory organizations with smart grid activities of interest and varied locations, sizes, and regulatory structures. For the third objective, we convened, with the assistance of the National Academy of Sciences, a panel of 23 experts in smart grid cybersecurity, including experts from utilities, vendors, manufacturers, researchers, and trade associations.

We conducted this performance audit from November 2009 to January 2011, in accordance with generally accepted government auditing standards. Those standards require that we plan and perform the audit to obtain sufficient, appropriate evidence to provide a reasonable basis for our findings and conclusions based on our audit objectives. We believe that the evidence obtained provides a reasonable basis for our findings and conclusions based on our audit objectives. Appendix I contains additional details on our objectives, scope, and methodology.

BACKGROUND

Background

The electricity industry, as shown in figure 1, is composed of four distinct functions: generation, transmission, distribution, and system operations. Once electricity is generated—whether by burning fossil fuels; through nuclear fission; or by harnessing wind, solar, geothermal, or hydro energy—it is generally sent through high-voltage, high-capacity transmission lines to local electricity distributors. Once there, electricity is transformed into a lower voltage and sent through local distribution lines for consumption by industrial plants, commercial businesses, and residential consumers. Because electric

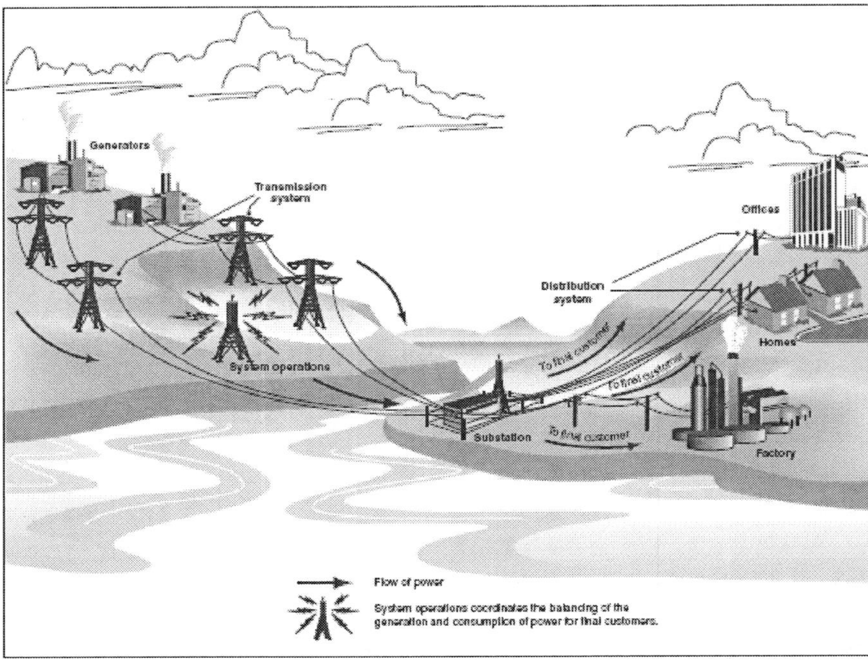

Source: GAO analysis.

Figure 1. Functions of the Electricity Industry.

energy is generated and consumed almost instantaneously, the operation of an electric power system requires that a system operator constantly balance the generation and consumption of power.

Utilities own and operate electricity assets, which may include generation plants, transmission lines, distribution lines, and substations—structures often seen in residential and commercial areas that contain technical equipment such as switches and transformers to ensure smooth, safe flow of current and regulate voltage. Utilities may be owned by investors, municipalities, and individuals (as in cooperative utilities). System operators—sometimes affiliated with a particular utility or sometimes independent and responsible for multiple utility areas—manage the electricity flows. These system operators manage and control the generation, transmission, and distribution of electric power using control systems—IT- and network-based-systems that monitor and control sensitive processes and physical functions, including opening and closing circuit breakers.[2] As we have previously reported, the effective functioning of the electricity industry is highly dependent on these control systems.[3] See the list of related past GAO products at the end of this report. However, for many years aspects of the electricity network lacked adequate technologies—such as sensors—to allow system operators to understand key information to detect how much electricity was flowing on distribution lines, communications networks to further integrate parts of the electricity grid with control centers, and computerized control devices to automate system management and recovery.

Smart Grid

As the electricity industry has matured and technology has advanced, utilities have begun taking steps to update the electricity grid—the transmission and distribution systems—by integrating new technologies and additional IT systems and networks. Though utilities have regularly taken such steps to upgrade their electricity systems, industry and government stakeholders have begun to articulate a broader, more integrated vision for transforming today's electricity grid into one that is more reliable and efficient, facilitates alternative forms of generation— including renewable energy, and gives consumers real-time information about fluctuating electricity costs.

This vision—commonly referred to as smart grid—would increase the use of IT systems and networks and two-way communication to automate actions that system operators formerly had to make manually. These efforts are designed to, among other things, improve transmission of electricity from power plants to consumers, provide grid operators with more information about conditions on the electricity system, integrate new and improved technologies into the grid, and allow consumers to receive more information about electricity prices and availability from the electricity system. Smart grid modernization is an ongoing process and initiatives have commonly involved installing advanced metering infrastructure (smart meters) on homes and commercial buildings that enable two-way communication between the utility and the customer. For example, FERC estimated advanced metering use in the United States at 4.7 percent in 2008, compared to 0.7 percent in 2006. Initiatives have also involved adding "smart" components to provide the system operator with more detailed data on the conditions of the transmission and distribution systems and better tools to observe the overall condition of the grid (called wide-area situational awareness). These include advanced, "smart" switches on the distribution system that communicate with each other to reroute electricity around a troubled line; and high-resolution, time

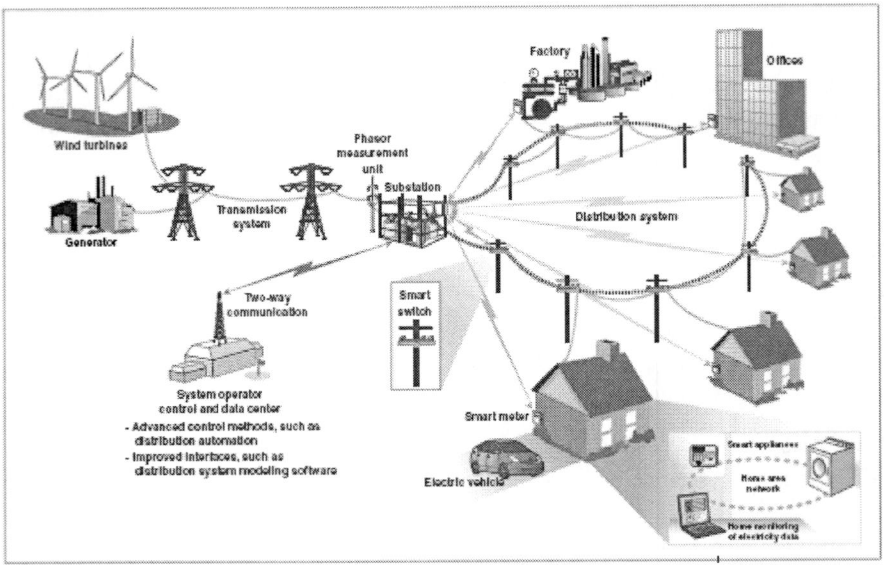

Source: GAO analysis.

Figure 2. Common Smart Grid Components.

synchronized monitors—called phasor measurement units—on the transmission system. Figure 2 illustrates one possible smart grid configuration. Utilities making actual smart grid investments may choose alternative configurations using different technologies and communications media depending on factors such as cost, customer needs, and local conditions.

Future smart grid applications may also include key roles for energy storage, in particular, storing electricity that is generated when it is inexpensive to produce. This may involve using improved battery technology, including the batteries in plug-in electric and hybrid-electric vehicles. Furthermore, smart grid systems may be used to encourage consumers to lower their demand for electricity during periods of high usage—called peaks. This could occur using home networks that automatically control appliances' electricity consumption in response to programmed consumer preferences and information about prices and demand received from the utility.[4]

According to the National Energy Technology Laboratory, a Department of Energy (DOE) national laboratory with a key role in supporting DOE smart grid efforts, smart grid systems fall into several different categories, as outlined in table 1.

Table 1. Categories of Smart Grid Systems as Defined by the National Energy Technology Laboratory

System category	Definition	Examples of related smart grid devices
Integrated communications	High-speed, fully integrated, two-way communication technologies that make the smart grid a dynamic, interactive "mega-infrastructure" for real-time information and power exchange. An open architecture facilitates an environment in which technologies from multiple vendors can easily interact and that securely connects grid components, customers, and operators, enabling them to talk, listen, and interact.	• Broadband over power line communications technologies • Wireless communications technologies such as WiFi • Home Area Networks—networks of appliances and other devices in the home
Advanced components	Advanced components that play an active role in determining the electrical behavior of the grid. These power system devices apply the latest research in materials, superconductivity, energy storage, power electronics, and microelectronics to produce higher power densities, greater reliability and power quality, enhanced electrical efficiency that produces major environmental gains, and improved real-time diagnostics.	• Advanced or "smart" switches, transformers, cables, and other electricaldevices • Storage devices, including plug-in hybrid electric vehicles, as well as advanced batteries • Grid-friendly, "smart" appliances, including air conditioners, clothes washers and dryers, and hot water heaters capable of delaying operation in response to price signals • Microgrids—local electricity grids that can

Table 1. (Continued)

System category	Definition	Examples of related smart grid devices
		operate independently of the main electricity grid when needed
Advanced control methods	• New methods and algorithms that monitor power system components, enabling rapid diagnosis and timely, appropriate response to any event. Integrating this information into planning models could improve utilization of generation and transmission assets.	• Substation and distribution automation—real-time monitoring and control of substation and distribution equipment • Fault locator systems that use sensors and digital information to locate faults—failures or drastic changes in current flowor total interruption of an electrical circuit
Sensing and measurement	Technologies that enhance power system measurements and enable the transformation of data into information. These technologies evaluate the health of equipment and the integrity of the grid, among other things. Such information enables consumers to make choices about whether to use electricity in response to information about electricity prices and demand, and can help provide relief when transmission lines are operating at or near capacity.	• Advanced sensors • Advanced metering infrastructure, including "smart" meters • Phasor measurement units—monitors that sample voltage and current many times a second at a given location on theelectricity grid to indicate grid stress and trigger corrective actions to maintain reliability • Dynamic line-rating devices that determine the real-time capacity of electrical lines • Consumer portals that provide consumers with real-time information about energy

System category	Definition	Examples of related smart grid devices
		consumption and prices
Improved interfaces and decision support	Decision support and improved interfaces that will enable more accurate and timely human decision making at all levels of the grid, including the consumer level, while also enabling more advanced operator training.	• Software tools to analyze the health of the electricity system • Distribution system modeling software • Real-time digital simulators to study and test electricity systems • Geographic information systems

Source: National Energy Technology Laboratory, A Compendium of Smart Grid Technologies. July 2009.

The use of smart grid systems may have a number of benefits, including improved reliability from fewer and shorter outages, downward pressure on electricity rates due to the ability to shift peak demand, an improved ability to transmit power from alternative energy sources such as wind, and an improved ability to detect and respond to potential attacks on the grid. It could also help consumers make more informed choices about when to use electricity; for example, how much to use when demand and prices are high. On the other hand, upgrading the grid would require major investments whose costs would ultimately be passed to utility consumers. Some electricity stakeholders, particularly those representing consumers, question whether the benefits of smart grid investments would be fully realized and have suggested that less costly approaches could achieve similar benefits.

State utility regulators are to evaluate applications for smart grid investments on a case-by-case basis. A number of these regulators have approved specific smart grid investments after determining that their benefits to consumers outweigh their costs.

According to the FERC-proposed smart grid policy statement, to achieve the smart grid characteristics and functions outlined in EISA, it is essential that these systems be interoperable—able to work with each other without special effort on the part of the customer. NIST officials explained that the electricity grid has historically relied on proprietary technology which is difficult to integrate with the technology of other manufacturers. In the case of smart grid upgrades, utilities have sought devices and systems that are interoperable and easily integrated with technologies from different vendors.

Smart Grid Cybersecurity

The smart grid vision and its increased reliance on IT systems and networks expose the electric grid to potential and known cybersecurity vulnerabilities associated with using such systems, which in turn increase the risk to the smooth and reliable operation of the electricity grid. As we and others have previously reported,[5] these potential vulnerabilities include:

- increasing the use of systems and networks increases the number of entry points and paths that can be exploited by potential adversaries and other unauthorized users;
- increasing the use of new system and network technologies can introduce new, unknown vulnerabilities;

- interconnecting systems and networks can allow adversaries wider access and the ability to spread malicious activity; and
- increasing the amount of customer information being collected on systems (and transmitting it via networks) provides monetary incentive for adversaries to attack these systems, and could lead to the unauthorized disclosure and use of private information.

In addition to these potential vulnerabilities, we and others have also reported that smart grid and related systems have known cyber vulnerabilities. For example, cybersecurity experts have demonstrated that certain smart meters can be successfully attacked, and the impact of such attacks includes the ability to disrupt the electricity grid. In addition, we reported in 2007 that certain smart systems—commonly referred to as control systems—used in industrial settings such as electric generation have cybersecurity vulnerabilities that, if exploited, could result in serious damages and disruption.[6] Further, in 2009, the Department of Homeland Security, in cooperation with a DOE national laboratory, ran a test that demonstrated that a vulnerability, commonly referred to as "Aurora," had the potential to allow unauthorized users to remotely control, misuse, and cause damage to a small commercial electric generator. Moreover, in 2008, the Central Intelligence Agency reported that malicious activities against IT systems and networks have caused disruption of electric power capabilities in multiple regions overseas, including a case that resulted in a multicity power outage.[7]

Smart Grid Regulation

Both the federal government and state governments have authority for overseeing the electricity industry. With respect to the electricity prices and rates of investor-owned utilities, wholesale electricity sales and transmission of electricity in interstate commerce are regulated by the federal government, specifically FERC.[8] This involves approving whether to allow utilities to recover the costs of investments they make to the transmission system. State public utility commissions (PUC) generally have authority to regulate local distribution and retail sales of electricity by investor-owned utilities in their state, including whether to allow these utilities to recover the costs of investments made to the distribution system. For cooperative and some municipal utilities, whose rate regulation by FERC and many state public utility commissions is limited, municipal city councils or cooperative boards of directors will generally approve cost recovery for electric investments. With respect to smart grid initiatives, individual utilities can choose to invest in smart grid devices on their own. However, as noted above, depending on the type of utility and where cost recovery is sought, either FERC, the state PUC, or another entity will have authority for deciding whether to allow that utility to recover the costs of smart grid investments from customers.

State and federal authorities also play key roles with respect to reliability, which can be affected by a system's cybersecurity. State regulators generally have authority to oversee the reliability of the local distribution system. The North American Electric Reliability Corporation (NERC) is the federally designated U.S. Electric Reliability Organization overseen by FERC. NERC has responsibility for conducting reliability assessments and enforcing mandatory standards to ensure the reliability of the bulk power system—a term that refers to facilities and control systems necessary for operating the electric transmission network and certain generation facilities needed for reliability. NERC develops reliability standards collaboratively through a deliberative process involving utilities and others in the electricity industry—which are then sent to FERC for approval.[9]

These reliability standards include critical infrastructure protection standards for protecting electric utility-critical and cyber-critical assets. In 2008, FERC approved eight critical infrastructure standards developed by NERC. These standards established requirements to help ensure the secure electronic exchange of information needed to operate and support the reliability of the bulk power system, and to help prevent unauthorized physical or electronic access to critical cyber assets. The eight standards require certain users, owners, and operators of the bulk power system to establish policies, plans, and procedures to safeguard physical and electronic access to control systems; identify and protect critical cyber assets; train personnel on security matters; report security incidents; and be prepared to recover from a cyber incident. NERC staff is engaged in the NIST-facilitated process, in particular, to address whether new or modified reliability standards will be necessary to ensure the continued reliability of the bulk power system as new smart grid technologies and systems are developed and integrated with existing systems and networks.

Recent Federal Smart Grid Activities

In 2007, EISA established that it is federal policy to support the modernization of the electricity grid and required actions by a number of federal agencies, including NIST, FERC, and DOE.[10] Specifically, the act directed NIST's Director, who reports to the Secretary of Commerce, to coordinate development of a framework of, among other things, IT standards for achieving the interoperability of smart grid systems. To accomplish this, NIST, starting in 2009, facilitated a process with stakeholders (e.g., utilities, smart grid technology vendors, standards development organizations, and others) to identify interoperability and cybersecurity standards related to smart grid. In January 2010, NIST reported that this process resulted in the identification of 75 standards that support smart grid interoperability. Of these, 11 involved cybersecurity.[11]

In addition to the NIST efforts to develop a framework for identifying interoperability and cybersecurity standards, the agency also identified the need to institute an initiative to develop cybersecurity guidelines for organizations such as electric companies, IT system vendors, and others involved in developing and implementing smart grid systems.

To carry out the above tasks (i.e., developing the standards framework and drafting the cybersecurity guidelines), NIST planned to establish two key working groups that are described in table 2.

With regard to FERC, under EISA the commission is to adopt those standards (identified as part of the NIST efforts) that it deemed necessary to ensure smart grid systems operate as intended. The act calls for FERC to institute a rule-making proceeding to accomplish this.[12]

Further, with regard to DOE, EISA authorized the department to establish two initiatives to facilitate development of industry smart grid efforts—the Smart Grid Investment Grant Program and the Smart Grid Regional Demonstration Initiative. DOE made $3.5 billion and $685 million of American Recovery and Reinvestment Act (Recovery Act) [13] funds available for these respective initiatives. In October 2009, under the Smart Grid Investment Grant Program, DOE announced awards for 100 grants to utilities in multiple states to stimulate the rapid deployment and integration of advanced digital technology needed to modernize the nation's electric grid. In November 2009, under DOE's Smart Grid Regional Demonstration Initiative, the department announced awards for 32 grants to fund regional demonstrations to verify technology viability, quantify costs and benefits, and validate new business models for the smart grid at a scale that can be readily adopted around the country.

Table 2. Key NIST Smart Grid Working Groups

Group name	Description
Smart Grid Interoperability Panel	A public-private partnership—which was initiated by NIST in 2009— to carry out a variety of tasks related to the development of a smart grid framework for interoperability and cybersecurity standards. This included • prioritizing and coordinating smart grid standards developed by stakeholders in the NIST process; and • administering priority action plans that identify where a new standard or extension of a standard is needed.
Smart Grid Cyber Security Working Group	A permanent working group of the Smart Grid Interoperability Panel that is to provide expertise needed to address matters related to smart grid cybersecurity. Among other things, this group is to be responsible for • developing smart grid cybersecurity guidelines; and • determining if the NIST-identified smart grid standards adequately address cybersecurity, including aligning with these guidelines.

Source: GAO analysis of NIST documents.

In addition to these recent actions, the federal government has undertaken other initiatives to facilitate the implementation of industry smart grid efforts, including funding technical research and development, data collection, and coordination activities (for more details on these efforts see appendix III). Most of these initiatives have been led by DOE.

NIST HAS DEVELOPED AND ISSUED SMART GRID CYBERSECURITY GUIDELINES, BUT THEY DO NOT ADDRESS SOME KEY CYBERSECURITY ELEMENTS

NIST developed, and issued in August 2010, a first version of its smart grid cybersecurity guidelines. To do this, NIST established in March 2009, the smart grid cyber security working group[14] to, among other things, develop guidelines for entities (e.g., utilities, equipment manufacturers, and regulators) to secure their smart grid systems. NIST intended the guidelines to, among other things, provide a process for entities to follow for developing solutions to address the security of their smart grid systems. To develop the guidelines, NIST planned to have the working group perform an assessment of the cybersecurity risks associated with existing and planned smart grid systems and then use the risk information, and an assessment of the privacy implications of these systems, to identify security requirements (i.e., controls) essential to securing such systems. As part of this assessment, NIST planned to address other key elements of cybersecurity, including the impact of coordinated cyber-physical attacks,[15] and identifying smart grid system vulnerabilities. The working group intended to complete these efforts and issue the guidelines in June 2010.

The working group has largely completed these steps, including issuing the guidelines. Specifically, during 2009 and 2010, the working group defined and then performed a high-level risk assessment of existing and planned smart grid systems—such as for transporting and storing electricity, and for advanced metering infrastructure. The risk assessment included identifying assets, vulnerabilities, and threats as well as specifying impacts for these and other systems as a means to identify security requirements (i.e., controls)—such as access control policies and procedures, employee training programs, incident response, and risk management—for securing such systems.

Using the results of the risk assessment and other efforts, the working group issued the smart grid cybersecurity guidelines in August 2010.[16] The guidelines include important elements, such as a high-level strategy that organizations can use to develop an approach to securing their smart grid systems, including identifying appropriate security requirements. In addition, the guidelines

- identified potential cryptography[17] issues that entities may encounter and solutions for resolving these issues;
- included a privacy impact assessment for the smart grid with a discussion of mitigating factors;
- identified potential smart grid vulnerabilities, as well as the possible impacts to organizations should the vulnerabilities be exploited;
- identified smart grid security problems, including how to ensure that access can be gained to critical devices and systems by personnel when ordinary authentication fails for any reason, and how to ensure that updates utilities send to smart meters are secure;
- detailed cybersecurity design issues, such as for password complexity rules; and
- identified smart grid cybersecurity areas requiring further research and development.

NIST stated in the guidelines that this initial version was to be updated periodically to incorporate any emerging issues.

While NIST largely addressed the key elements in developing its guidelines, it did not address an important element essential to securing smart grid systems and networks that NIST had planned to include. Specifically, it did not address the risk of combined cyber-physical attacks.

NIST also identified other key elements that surfaced during its development of the guidelines that need to be addressed in future guideline updates. These include identifying

- research and development that needs to be performed, such as for synchrophasor[18] security;
- cryptography issues, and solutions to resolve cryptography issues; and
- additional smart grid system design issues, such as managing vulnerabilities incurred in the supply chain.

NIST officials said they did not address the cyber-physical and other above topics in the guidelines because, in part, they had not yet fully developed these sections by the planned June 2010 issuance date. Consequently, if NIST had taken the time to address and incorporate these topics, it would have caused the agency to have been even further behind schedule, meaning the guidelines would have been issued later than August 2010.

NIST officials also said that the working group intends to update the guidelines to, among other things, address these missing elements To do so, NIST drafted a plan and schedule for updating the cybersecurity guidelines periodically. While a positive step, the plan and schedule, as of October 2010, were still in draft form. NIST officials stated that they are in the process of rewriting the plan and schedule and intend to have them finalized by the end of the year.

Having a finalized plan and schedule with specific milestones is critical for ensuring the guidelines fully address key cybersecurity elements that have not been incorporated thus far. Without it, there is increased risk that important cybersecurity elements will not be addressed by entities implementing smart grid systems, thus making these systems vulnerable to attack.

FERC Has Begun Reviewing Initial Smart Grid Standards but Has Not Developed a Coordinated Plan to Monitor Industry's Implementation

In 2010, FERC began reviewing for adoption an initial set of smart grid interoperability and cybersecurity standards developed through the NIST standards process. However, FERC has not developed a coordinated approach with other regulators to monitor the extent to which industry follows these voluntary standards, because, according to officials, it has not yet determined whether or how to perform such a task. Without a documented approach to coordinate with state and other regulators on this issue, FERC will not be well positioned to promptly begin monitoring the results of any standards it adopts or quickly respond if gaps arise.

FERC Has Begun Reviewing an Initial Set of Smart Grid Standards, but Enforcement Authorities Are Divided among Multiple Regulators

In October 2010, FERC began its process of reviewing for adoption smart grid standards related to interoperability and cybersecurity, but authority to enforce these standards is divided among multiple regulators. The five standards being initially reviewed were identified by NIST as ready for regulator consideration and represent a subset of those identified through the NIST-facilitated smart grid standards process.[19] FERC designated a docket for a proceeding to review these five standards and adopt those that it believes are necessary to ensure smart grid functionality and interoperability in interstate transmission of electric power and regional and wholesale electricity markets. FERC staff were uncertain when the initial set of standards would be adopted, but both FERC and NIST officials told us that, because smart grid standards are continually evolving, they expect multiple rounds of standards to be reviewed and adopted by FERC. FERC staff have suggested various criteria that they believe the Commissioners should use when considering whether to adopt the standards, including recommending relying on the assessment of the NIST Cyber Security Working Group and rule making comments to determine if cybersecurity has been adequately incorporated. FERC also provided guidance to help NIST prioritize interoperability standards development. In a July 2009 smart grid policy statement, FERC proposed prioritizing two crosscutting issues—system security (including cybersecurity) and intersystem communication—along with four key functionalities—wide area situational awareness,[20] demand response, electric storage, and electric transportation.

While EISA gives FERC authority to adopt smart grid standards, it does not provide FERC with specific enforcement authority. In particular, EISA gives FERC the authority to adopt standards once it finds the NIST process has led to sufficient consensus. However, according to FERC officials, the statute did not provide specific additional authority to allow FERC to require utilities or manufacturers of smart grid technologies to follow these standards. As a result, any standards identified and developed through the NIST-led process are voluntary unless regulators use other authorities to indirectly compel utilities and manufacturers to follow them. Stakeholders we spoke with—federal electricity officials, participants in the smart grid standards development process, and other electricity and cybersecurity experts—noted that, while voluntary industry-developed standards have historically been used in the electricity industry, some factors could limit the extent to which they are followed. Although some explained that economic and market pressure should encourage manufacturers and utilities to follow voluntary standards, others noted that there could still be gaps in the extent to which the standards are followed, particularly if the cost of following standards is high or if utilities have varying levels of familiarity with and interest in implementing them.

According to FERC officials, FERC's only authority to require utilities to follow standards or use standards-compliant devices would derive from its existing reliability and cost-recovery authorities under the Federal Power Act, which generally apply to transmission assets.[21] For example, FERC could require that utilities subject to its rate regulation use standards-compliant smart grid devices as a condition of allowing them to recover the costs of smart grid investments on the transmission system. Additionally, to the extent that interoperability and cybersecurity standards are deemed necessary to ensure the reliability of the bulk power system, such standards could be considered through the NERC standards-setting process, and if approved, would be considered mandatory and enforceable by both NERC and FERC. However, FERC officials noted that NERC's reliability standards-setting process involves extensive deliberation by industry; that it is possible that NERC could choose not to develop a mandatory reliability standard that FERC had adopted through its separate process for smart grid standards; and that FERC is prohibited from adopting reliability standards on its own outside of the NERC process.

The fragmented nature of electricity industry regulation further complicates enforcement of smart grid standards and oversight of smart grid investments using FERC and other regulators' existing authorities.[22] Oversight responsibility is divided among various regulators at the federal, state, and local level, and FERC's authority is limited to certain parts of the grid, generally the transmission system. As a result, state regulatory bodies and other regulators with authority over the distribution system will play a key role in overseeing the extent to which interoperability and cybersecurity standards are followed since many smart grid upgrades will be installed on the distribution system. Such regulatory fragmentation can make it difficult for individual regulators to develop an industry-wide understanding of whether utilities and manufacturers are following voluntary standards. This is due to the large number of regulators in the industry—FERC, electricity regulators in 50 states and the District of Columbia, and regulators of thousands of cooperative and municipal utilities—and their potentially limited visibility over parts of the grid outside their jurisdiction.

The state public utility commissions we spoke with were at different points in developing their approach to monitoring smart grid interoperability and cybersecurity.[23] Multiple state regulators told us that, while they have not imposed any formal requirements on utilities with respect to the interoperability and cybersecurity of smart grid technologies, their offices have ongoing conversations with regulated utilities about the issue. Others have established requirements in PUC rule makings outlining minimum functionalities that smart meters must achieve, and in the case of the Public Utility Commission of Texas, audits that smart meter manufacturers must obtain to demonstrate that smart meter data can be securely accessed by customers and others. Additionally, the California and Colorado commissions have opened proceedings to initiate discussion with the public about how to best address topics like the interoperability and cybersecurity of smart grid technologies. Finally, most PUC staff were uncertain what approach their Commissions would take to enforce any standards that FERC decides to adopt, and three said that limited resources and technical expertise made their roles in overseeing interoperability and cybersecurity, including participating in the NIST standards process, more challenging.

A number of cooperatively and municipally owned utilities fall partially out of the purview of federal and state regulators, and as such, it will be up to their regulators—often utility boards of directors—to oversee the interoperability and cybersecurity of their smart grid efforts. In Nebraska, for example, the state is entirely composed of consumer owned utilities, including municipal and cooperative utilities and public power districts. This means that, in part, oversight of smart grid interoperability and cybersecurity in this state will fall to the numerous individual regulators of these utilities. In addition, there are thousands of cooperatively and municipally owned utilities located across the country.

FERC Has Not Developed a Coordinated Approach to Monitor Whether Industry Follows Voluntary Standards

Despite the importance of ensuring manufacturers and utilities follow smart grid standards, FERC has not developed an approach coordinated with other regulators to monitor at a high level the extent to which industry will follow the voluntary smart grid standards it adopts. There have been some initial efforts by regulators to share views. For example, a collaborative dialogue between FERC and the National Association of Regulatory Utility

Commissioners (NARUC) to facilitate the transition to a smart electric grid—the FERC-NARUC Collaborative on Smart Response—has discussed the standards-setting process in general terms. However, according to FERC and NARUC officials, FERC and the state PUCs have not established a joint approach for monitoring how widely voluntary smart grid standards are followed in the electricity industry or developed strategies for addressing any gaps. According to FERC officials and others representing municipal and cooperative utilities, FERC also has not coordinated in such a way with groups representing public power and cooperative utilities—utilities not routinely subject to FERC's or the states' jurisdiction for rate-setting purposes. Such groups include the American Public Power Association, which represents municipally owned utilities, and the National Rural Electric Cooperative Association, which represents cooperatively owned utilities. FERC has not developed such an approach, because, according to officials, it has not yet determined whether or how to conduct high-level monitoring of compliance with smart grid standards it adopts under EISA

Adherence to standards is an important step toward achieving an interoperable and secure electricity system. Unless FERC and other regulators have a good understanding of whether utilities and manufacturers are following smart grid standards, it will be difficult to know whether a voluntary approach to standards setting is effective or if changes are needed. According to federal internal control guidance, managers need to compare actual performance—in this case, the extent to which manufacturers and utilities follow voluntary standards—to planned or expected results throughout the organization and analyze significant differences.[24] Given the fragmented nature of electricity regulation, it may not be possible for FERC to perform such a review alone, and the agency may have to collaborate with other regulators. Past GAO work highlights that when carrying out activities to enhance interagency collaboration, it is critical to involve nonfederal partners—in this case, state and other regulators with responsibility for overseeing key components of the electricity industry—in decision making.[25] Without a documented approach established in advance to coordinate with state and other regulators on this issue, FERC will not be well positioned to promptly begin monitoring the results of any standards it adopts—including a high-level assessment of whether industry follows them—and quickly respond if gaps arise. Such a delay could result in a patchwork of approaches across the United States and lead to incompatibilities between systems, higher costs, and a less secure electricity grid.

A number of activities are under way that may result in information to inform a FERC assessment of the extent to which voluntary standards are followed, but these efforts are not coordinated or complete. According to DOE officials, as a part of DOE's broader effort to publish a smart grid system report every 2 years as required by EISA, the department expects to report some information about the progress and effectiveness of smart grid interoperability and cybersecurity standards. Additionally, NIST has efforts under way to establish a process for vendors to certify their smart grid products as complying with standards and coordinate industry development of additional standards as needed. However, it is unclear to what extent these planned activities will specifically focus on assessing industry compliance with voluntary standards across regulatory jurisdictions and options to address any gaps that exist. Moreover, unlike FERC, the state PUCs, and other electricity regulators, neither DOE nor NIST has the authority to routinely require industry to follow standards should gaps exist.

ELECTRICITY INDUSTRY FACES KEY CHALLENGES TO SECURING SMART GRID SYSTEMS AND NETWORKS

Leveraging the views of experts (by means of panel discussions), we identified the following six challenges that are key to ensuring the cybersecurity of the systems and networks that support our nation's electricity grid.

- *Aspects of the current regulatory environment make it difficult to ensure the cybersecurity of smart grid systems.* In particular, jurisdictional issues and the difficulties associated with responding to continually evolving cyber threats are a key regulatory challenge to ensuring the cybersecurity of smart grid systems as they are deployed. Regarding jurisdiction, our experts expressed concern that there was a lack of clarity about the division of responsibility between federal and state regulators, particularly regarding cybersecurity. While jurisdictional responsibility has historically been determined by whether a technology is located on the transmission or distribution system, experts raised concerns that smart grid technology may blur these lines. For example, devices such as smart meters deployed on parts of the grid traditionally subject to state jurisdiction could, in the aggregate, have an impact on those parts of the grid that federal regulators are responsible for—namely the reliability of the transmission system.

There is also concern about the ability of regulatory bodies to respond to evolving cybersecurity threats. For example, one expert questioned the ability of government agencies to adapt to rapidly evolving threats, while another highlighted the need for regulations to be capable of responding to the evolving cybersecurity issues. In addition, our experts expressed concern with agencies developing regulations in the future that are overly specific in their requirements, such as those specifying the use of a particular product or technology. Consequently, unless steps are taken to mitigate these challenges, regulations may not be fully effective in protecting smart grid technology from cybersecurity threats.

- *Consumers are not adequately informed about the benefits, costs, and risks associated with smart grid systems.* Specifically, there is concern that consumers are not aware of the benefits, costs, and risks associated with smart grid systems. This lack of awareness may limit the extent to which consumers are willing to pay for secure and reliable systems, which may cause regulators to be reluctant to approve rate increases associated with cybersecurity. As a result, until consumers are more informed about the benefits, costs, and risks of smart grid systems, utilities may not invest in, or get approval for, comprehensive security for smart grid systems, which may increase the risk of attacks succeeding.
- *Utilities are focusing on regulatory compliance instead of comprehensive security.* The existing federal and state regulatory environment creates a culture within the utility industry of focusing on compliance with cybersecurity requirements, instead of a culture focused on achieving comprehensive and effective cybersecurity. Specifically, experts told us that utilities focus on achieving minimum regulatory requirements rather than designing a comprehensive approach to system security. In addition, one expert stated that security requirements are inherently incomplete, and having a culture that views the security problem as being solved once those requirements are met will leave an organization vulnerable to cyber attack. Consequently, without a comprehensive approach to security, utilities leave themselves open to unnecessary risk.

- *There is a lack of security features being built into smart grid systems.* Security features are not consistently built into smart grid devices. For example, our experts told us that certain currently available smart meters have not been designed with a strong security architecture and lack important security features, including event logging[26] and forensics capabilities which are needed to detect and analyze attacks. In addition, our experts stated that smart grid home area networks—used for managing the electricity usage of appliances and other devices in the home—do not have adequate security built in, thus increasing their vulnerability to attack. Without securely designed smart grid systems, utilities will be at risk of not having the capacity to detect and analyze attacks, which increases the risk that attacks will succeed and utilities will be unable to prevent them from recurring.
- *The electricity industry does not have an effective mechanism for sharing information on cybersecurity and other issues.* The electricity industry lacks an effective mechanism to disclose information about smart grid cybersecurity vulnerabilities, incidents, threats, lessons learned, and best practices in the industry. For example, our experts stated that while the electricity industry has an information sharing center, it does not fully address these information needs. In addition, President Obama's cyberspace policy review, released in May 2009, also identified challenges related to cybersecurity information sharing within the electric and other critical infrastructure sectors and issued recommendations to address the areas.[27] According to our experts, information regarding incidents such as both unsuccessful and successful attacks must be able to be shared in a safe and secure way to avoid publicly revealing the reported organization and penalizing entities actively engaged in corrective action. Such information sharing across the industry could provide important information regarding the level of attempted cyber attacks and their methods, which could help grid operators better defend against them. If the industry pursued this end, it could draw upon the practices and approaches of other industries when designing an industry-led approach to cybersecurity information sharing. Without quality processes for information sharing, utilities will not have the information needed to adequately protect their assets against attackers.
- *The electricity industry does not have metrics for evaluating cybersecurity.* The electricity industry is also challenged by a lack of cybersecurity metrics, making it difficult to measure the extent to

which investments in cybersecurity improve the security of smart grid systems. Experts noted that while such metrics[28] are difficult to develop, they could help compare the effectiveness of competing solutions and determine what mix of solutions combine to make the most secure system. Furthermore, our experts said that having metrics would help utilities develop a business case for cybersecurity by helping to show the return on a particular investment. Until such metrics are developed, there is increased risk that utilities will not invest in security in a cost-effective manner, or have the information needed to make informed decisions on their cybersecurity investments.

CONCLUSION

The electricity industry is in the midst of a major transformation as a result of smart grid initiatives, and this transformation has led to significant financial investment by many entities, including utilities, private companies, and the federal government. For their part, NIST and FERC have efforts planned and under way to carry out their smart grid roles and responsibilities, although limitations exist in the planning and coordination efforts of these two key agencies. Specifically, NIST does not have a definitive plan and schedule, including specific milestones, for updating and maintaining its cybersecurity guidelines to address key missing elements. Furthermore, FERC has not established an approach coordinated with other regulators to monitor the extent to which industry is following the smart grid standards it adopts.

The voluntary standards and guidelines developed through the NIST and FERC processes offer promise. However, a voluntary approach poses some risks when applied to smart grid investments, particularly given the fragmented nature of regulatory authority over the electricity industry. Currently, NIST and FERC's efforts are hindered by their lack of an approach to (1) updating voluntary cybersecurity guidelines and (2) monitoring whether voluntary standards are being followed by manufacturers and utilities and periodically reporting to Congress on whether additional authorities are needed. Not having such an approach could result in gaps being recognized too late to avoid incompatibilities between systems, costly equipment replacements, or unnecessarily long periods of vulnerability to cyber attack. The lack of an approach to monitoring compliance with standards also limits the information available to Congress on how widely the smart grid standards are being followed and whether additional regulatory authorities are needed to address any gaps.

In addition to the challenges being faced by NIST and FERC, the electricity industry faces its own set of challenges that are critical to ensuring smart grid systems and networks are implemented securely. Addressing these challenges will involve participation by private sector organizations and government agencies, including NIST and FERC. Because these two agencies are key to addressing the challenges, it is especially important that NIST and FERC when addressing their planning and coordination limitations also consider whether the challenges should be addressed in their current and planned cybersecurity efforts.

RECOMMENDATIONS FOR EXECUTIVE ACTION

To reduce the risk that NIST's smart grid cybersecurity guidelines will not be as effective as intended, we recommend that the Secretary of Commerce direct the Director of NIST to finalize the agency's plan for updating and maintaining the cybersecurity guidelines, including ensuring it incorporates (1) missing key elements identified in this report, and (2) specific milestones for when efforts are to be completed. We also recommend that NIST, as a part of finalizing the plan, assess whether any cybersecurity challenges identified in this report should be addressed in the guidelines.

To improve coordination among regulators and help Congress better assess the effectiveness of the voluntary smart grid standards process, we recommend that the Chairman of FERC, making use of existing smart grid information, develop an approach to

- coordinate with state regulators to (1) periodically evaluate the extent to which utilities and manufacturers are following voluntary interoperability and cybersecurity standards and (2) develop strategies for addressing any gaps in compliance with standards that are identified as a result of this evaluation. To the extent that FERC determines it lacks authority to address any gaps in compliance that cannot be addressed through this coordinated approach with other regulators, the Chairman should report this information to Congress.
- coordinate with groups that represent utilities subject to less FERC and state regulation (such as municipal and cooperative utilities) to (1)

> periodically evaluate the extent to which utilities and manufacturers are following voluntary interoperability and cybersecurity standards and (2) develop strategies for addressing any gaps in compliance with standards that are identified as a result of this evaluation. To the extent that FERC determines it lacks authority to address any gaps in compliance that cannot be addressed through this coordinated approach, the Chairman should report this information to Congress.

We also recommend that the Chairman of FERC, working with NERC as appropriate, assess whether any cybersecurity challenges identified in this report should be addressed in commission cybersecurity efforts.

AGENCY COMMENTS AND OUR EVALUATION

In written comments—signed by the Secretary of Commerce and the Chairman of FERC (see appendixes IV and V, respectively)—on a draft of this report, both agencies stated that they agreed with our recommendations.

Although Commerce agreed with the recommendations, the department (1) offered three related comments on a finding in the report, and (2) suggested rewording part of our recommendations based on those comments. Specifically, in the first two comments, the department wanted to replace wording we used in the report (i.e., replacing "missing key elements" with "NIST's follow-on cyber-physical activity") and delete two report sentences which, in its view, incorrectly implied that NIST planned to complete its cyber-physical activity and report its work results in the issuance of the August 2010 guidelines. In its third comment, Commerce agreed that the risk of combined cyber-physical attacks needs to be addressed in the guidelines, but reiterated its disagreement with our report statement that NIST was planning to cover this in the August 2010 guidelines. Based on these comments, the department suggested wording changes to part of our recommendations to reflect its view. However, our review of drafts of the guidelines, including one issued by NIST to the public in February 2010, coupled with discussions with NIST officials responsible for developing the guidelines, show that that the agency had planned to address this topic in the August 2010 version of the guidelines. Based on this evidence, we did not make any changes to our report.

In addition to agreeing to our recommendations, FERC also (1) commended the draft report's discussion of cybersecurity for the electric industry, (2) said it appreciated the report's conclusions, and (3) described steps it intended to take to implement the recommendations. Specifically, with regard to our recommendation to improve coordination among regulators, FERC stated that it intends to direct commission staff to evaluate possible approaches to improving coordination among regulators. In addition, FERC stated that if the commission finds that it lacks authority to address gaps in electric industry compliance with voluntary interoperability and cybersecurity standards, it intends to report this information to Congress as our report recommends. Further, in response to our recommendation to assess whether any of the challenges identified in our report should be addressed in commission cybersecurity efforts, FERC said it had directed commission staff to develop procedures to perform such an assessment.

In addition to the above comments, FERC also presented two general issues with the report. The first is that while FERC agreed with the challenge associated with the lack of cybersecurity metrics, identified in the draft, it commented that developing valid metrics also presents a separate challenge of its own. We agree with this view and believe it is consistent with our report findings. The second issue is that according to FERC, our report appeared to assume that all relevant manufacturers and utilities are to comply with the voluntary standards being developed through the process specified in EISA. To clarify, we neither stated this assumption in our report nor was it our intent to imply such an assumption. Nonetheless, it is important to note that the findings described in our report show it is critical for FERC to determine the extent to which these standards are being followed, and that is why we included a recommendation for the agency to coordinate with state regulators and others to achieve this goal.

We also provided a copy of the draft report for review and comment to DOE. In an e-mail from the Team Lead for Strategic Planning and Daily Operations within DOE's Office of Electricity Delivery and Energy Reliability, the department provided technical comments on the report, which we incorporated as appropriate.

David A. Powner
Director,
Information Technology Management Issues

David C. Trimble
Acting Director,
Natural Resources and Environment

APPENDIX I: OBJECTIVES, SCOPE AND METHODOLOGY

Our objectives were to (1) assess the extent to which the National Institute of Standards and Technology (NIST) has developed smart grid cybersecurity guidelines, (2) evaluate the Federal Energy Regulatory Commission's (FERC) efforts to adopt smart grid cybersecurity and other standards and monitor their use by industry, and (3) identify challenges associated with ensuring the cybersecurity of the smart grid.

For our first objective, we analyzed applicable laws to determine NIST's responsibilities with respect to the smart grid. Then we analyzed agency plans and related documentation and interviewed responsible officials to determine the steps NIST was planning to take or had taken to meet those responsibilities. Specifically, we analyzed NIST's plans for developing smart grid cybersecurity guidelines, and compared them with the issued guidelines[1] to identify any differences. Where there was a difference between NIST's plans and what had been completed, we analyzed the impact of the difference and its cause.

For the second objective, we analyzed FERC documentation, including their interim and final Smart Grid Policy Statement,[2] and reviewed relevant laws and regulations. We interviewed FERC staff to better understand their authority with respect to smart grid standards, expected approach to standards adoption, and the extent of coordination with other regulators. We also interviewed state electricity regulators to understand their regulatory approach and perspectives on smart grid standards b eing identified and developed through the NIST process. The state regulators we sought the views of included the Alabama Public Service Commission, California Public Utilities Commission, Colorado Public Utilities Commission, Nebraska Power Review Board, Public Utilities Commission of Ohio, Pennsylvania Public Utility Commission, and the Public Utility Commission of Texas. These states were selected because they had smart grid activities of interest and were generally varied in terms of location, size, and regulatory structure. As part of this work, we identified the steps taken by these states to oversee interoperability and cybersecurity of smart grid investments, although we did not evaluate their adequacy. In Nebraska, because all utilities are consumer-owned, state electricity regulators do not have authority to oversee whether smart grid investments are interoperable or cyber secure. As a result, we excluded Nebraska from any summaries of state responses presented in the body of this report. We also met with staff from two groups representing public and cooperatively owned utilities: the American Public Power Association and the

National Rural Electric Cooperative Association. Additionally, we reached out to various electricity experts, including representatives of standards development organizations, participants in the NIST standards development process, and others, to gather their opinions on the strengths and limitations of the NIST approach and standards setting.

For our third objective, we convened a panel of experts in coordination with the National Academy of Sciences. Specifically, we worked iteratively with the National Academy of Sciences' Computer Science and Telecommunication Board to choose a group of panel members with expertise in subjects most applicable to our objective. The selected experts included representatives from electric utilities responsible for implementing and securing smart grid systems, public utility commissions, trade associations, smart grid technology vendors, and cybersecurity experts. A full list of the expert panelists can be found in appendix III. A key topic discussed by the panel was the major cybersecurity challenges facing the grid, and related issues, such as the potential consequences of security failures, adequacy of current cybersecurity technology, effectiveness of regulatory frameworks and enforcement mechanisms, potential benefits for key stakeholder groups, and additional steps regulators could take to ensure that smart grid investments are secure. We then analyzed the results of the panel, and from that analysis developed a list of the major challenges and a summary of each. We then had the panelists review the list and our accompanying summary to make sure we accurately captured their views.

We conducted this performance audit from November 2009 to January 2011, in accordance with generally accepted government auditing standards. Those standards require that we plan and perform the audit to obtain sufficient, appropriate evidence to provide a reasonable basis for our findings and conclusions based on our audit objectives. We believe that the evidence obtained provides a reasonable basis for our findings and conclusions based on our audit objectives.

APPENDIX II: OTHER FEDERAL EFFORTS TO FACILITATE SMART GRID IMPLEMENTATION

In addition to efforts discussed in this report, the federal government has undertaken other efforts to facilitate smart grid implementation, including conducting and funding technical research and development, data collection, and coordination activities. Most of these initiatives have been led by the Department of Energy (DOE). Table 3 describes 10 of these other key efforts, including the federal agency involved and the purpose of the effort.

Table 3. Other Federal Efforts to Support Smart Grid Implementation

Effort	Federal agency that established effort and year	Membership or contractors	Purpose of effort
Smart Grid Cybersecurity Web site	Established by DOE and its Pacific Northwest National Laboratory and launched in January 2010	DOE and Pacific Northwest National Laboratory	To share information about smart grid cybersecurity and cybersecurity approaches used in smart grid demonstration and investment grant projects.
GridWise Architecture Council	Established by DOE in 2004	Recognized practitioners and leaders from the electricity industry and related sectors, including state governments and private representatives, ranging from major corporations to independent systems operators and others	To promote and enable interoperability among the many entities that interact with the nation's electric power system. The council enlists industry involvement to articulate the goal of interoperability across the electric system; identify the concepts and architectures needed to make interoperability possible, and develop actionable steps to facilitate the interoperation of systems, devices, and institutions that encompass the nation's electric system.
Federal Energy Regulatory Commission and National Association of Regulatory Utility Commissioners (FERC-NARUC) Collaborative on Smart Response	The smart grid component of this collaborative was established by FERC and NARUC in 2008	NARUC and FERC	To provide a forum for federal and state regulators to discuss the smart grid and demand response policies, share best practices and technologies, and address issues that benefit from state and federal collaboration.
North American SynchroPhasor Initiative	Established in 2007 by DOE and the North American Electric Reliability Corporation (	DOE; NERC; North American electric utilities, vendors, and consultants; federal and private	To promote power system reliability and visibility through wide-area measurement and control using smart grid synchrophasor technology. Its mission is to encourage a robust,

Table 3. (Continued)

Effort	Federal agency that established effort and year	Membership or contractors	Purpose of effort
	NERC) along with electric utility companies and other organizations	researchers; and academics	widely available, and secure synchronized data measurement infrastructure for the interconnected North American electric power system with associated analysis and monitoring tools for improved reliability and better planning and operation. This effort builds upon prior related DOE efforts.
Smart Grid Advisory Committee	Established in 2008 by DOE, as required by the Energy Independence and Security Act of 2007 (EISA)	Members are selected by the DOE secretary from both private and nonfederal public sector stakeholders based on their experience, expertise, and ability to represent the full range of smart grid technologies and services	To advise relevant federal officials concerning the development of smart grid technologies, the progress of a national transition to the use of smart grid technologies and services, the evolution of widely accepted technical and practical standards and protocols to allow interoperability and intercommunication among smart grid-capable devices, and the optimum means of using federal funding to encourage such progress. According to DOE officials, the committee was incorporated into the Electricity Advisory Committee.
Smart Grid Data Hub and smartgrid.gov Web site	Initiated by DOE in 2009. DOE contracted with the National Renewable Energy Laboratory to lead multiple labs in developing, establishing, and maintaining the data hub and Web site	DOE, the National Renewable Energy Laboratory, Lawrence Berkeley National Laboratory, Oak Ridge National Laboratory, and Navigant Consulting Inc.	To collect and maintain information about the American Recovery and Reinvestment Act-funded smart grid activities and their progress to help inform and educate consumers about all aspects of the smart grid.
Smart Grid Information Clearinghouse	Initiated by DOE in 2009. DOE contracted with Virginia Tech to design, establish,	DOE, Virginia Tech, the IEEE Power & Energy Society, and EnerNex	To serve as a repository for government and industry smart grid information including standards, projects, lessons learned, and best practices to

Effort	Federal agency that established effort and year	Membership or contractors	Purpose of effort
	and initially maintain this clearinghouse		facilitate wide-ranging data gathering and information sharing. It is located at http://www.sgiclearinghouse.org
Smart Grid Maturity Model	Originally developed by IBM and seven utilities. In 2009, Carnegie Mellon University's Software Engineering Institute became the steward for this model. DOE sponsors the institute's activities related to the model	DOE and Carnegie Mellon University	To provide a management tool to help utilities assess and improve their progress in implementing the smart grid.
Smart Grid Task Force	Established by DOE in 2008, as required by EISA	Electricity experts from DOE, FERC, the National Institute of Standards and Technology (NIST), and other federal entities (e.g., Environmental Protection Agency, Department of Homeland Security, Department of Agriculture, and Department of Defense)	To ensure awareness, coordination, and integration of federal government activities related to smart grid technologies, practices, and services.
DOE National Laboratory smart grid projects	Initiated over multiple years at various DOE labs	National Laboratories include: -Ames, -Argonne, -Lawrence Berkeley, -Idaho, - National Energy Technology	To support smart grid development by conducting and funding technical research and development projects, such as modeling; standards development and conformance testing; and development of smart grid devices and applications; among other things. For example, Argonne National Laboratory has a

Table 3. (Continued)

Effort	Federal agency that established effort and year	Membership or contractors	Purpose of effort
		Laboratory, -National Renewable Energy Laboratory, -Oak Ridge, -Pacific Northwest National Laboratory, and - Sandia National Laboratories	modeling project to analyze the impact of smart grid technologies (e.g., plug-in hybrid electric vehicles—vehicles that interconnect with the grid to charge and store electricity on grid infrastructure, electricity demand, and electricity prices). In addition, in the area of standards development, multiple labs participate in the NIST smart grid standards process. Additionally, several laboratories are developing smart grid devices and applications, such as Sandia, which is developing sensing, monitoring, and control devices to address the technical challenges associated with integrating renewable energy systems into the current transmission and distribution infrastructure.

Source: GAO analysis of agency and industry sources.

APPENDIX III: EXPERT PANEL DISCUSSION ATTENDEES

The names and affiliation of the experts who participated in the panel discussion held June 2-3, 2010, in Washington, D.C., are:

- Sharla Artz, Director of Government Affairs, Schweitzer Engineering Laboratories, Inc.
- David Baker, Director of Services, IOActive, Inc.
- David Batz, Manager, Cyber & Infrastructure Security, Edison Electric Institute
- W. Earl Boebert, Sandia National Laboratories (retired)
- Michael Butler, Senior Analyst, National Institute of Standards and Technology

- Matthew Carpenter, Senior Security Analyst, InGuardians
- Jeffrey E. Dagle, Chief Electrical Engineer, Energy Technology Development, Pacific Northwest National Laboratory
- David Dunn, Manager, Organizational Governance Support, Independent Electricity System Operator
- Robert Former, Principal Security Engineer, Itron, Inc.
- Travis Goodspeed, Security Consultant, Radiant Machines
- Ed Gray, Vice President, Legislative and Regulatory Affairs, Elster Solutions
- Donny Helm, Manager of Technology, Electric Delivery, Oncor
- Michael Hyland, Vice President of Engineering Services, American Public Power Association
- Stan M. Kaplan, Energy and Environmental Policy Specialist, Congressional Research Service
- Jeffrey S. Katz, Chief Technology Officer, Energy and Utilities Industry, IBM
- Christopher Knudsen, Director, Technology Innovation Center, Pacific Gas & Electric
- Stephen J. Lukasik, Former Director, Defense Advanced Research Projects Agency and Former Chief Scientist, Federal Communications Commission
- Richard Pethia, Director of the CERT program, Carnegie Mellon University Software Engineering Institute
- William H. Sanders, Donald Biggar Willett Professor of Engineering, Director, Information Trust Institute, and Acting Director, Coordinated Science Laboratory, University of Illinois
- Christopher Villarreal, Regulatory Analyst, California Public Utilities Commission
- David Wollman, Manager, Electrical Metrology Groups, National Institute of Standards and Technology
- Andrew Wright, Chief Technology Officer, N-Dimension Solutions
- Christine Wright, Team Leader, Competitive Markets Division, Public Utility Commission of Texas

Appendix IV: Comments from the Department of Commerce

UNITED STATES DEPARTMENT OF COMMERCE
The Secretary of Commerce
Washington, D.C. 20230

December 20, 2010

Mr. Gregory C. Wilshusen
Director, Information Security Issues
United States Government Accountability Office
Washington, DC 20548

Dear Mr. Wilshusen:

Thank you for the opportunity to comment on the draft report entitled "Electricity Grid Modernization: Progress Being Made on Cybersecurity Guidelines, but Key Challenges Remain to be Addressed" (GAO-11-117).

We agree with the report's recommendations that (1) the National Institute of Standards and Technology (NIST) finalize its plan and schedule for updating its cybersecurity guidelines, and (2) the Federal Energy Regulatory Commission develop a coordinated approach to monitor voluntary standards and address any gaps in compliance. We offer the following comments regarding GAO's conclusions.

1. Throughout the report, replace "missing key elements" with "NIST's follow-on cyber-physical activity."

2. Page 16. Delete the following two sentences in the first paragraph: "As a part of this assessment, NIST planned to address other key elements of cybersecurity, including the impact of coordinated cyber-physical attacks, and identifying smart grid system vulnerabilities. The working group intended to complete these efforts and issue the guidelines in June 2010." NIST stated that it plans to include these elements in revisions to the guidelines that were issued in September 2010, not those issued in June 2010. These sentences incorrectly imply that NIST has not completed planned activities related to cyber-physical attacks.

3. Page 29. NIST agrees that the risk of combined cyber-physical attacks on the smart grid is an area that needs to be more fully explored in the future, but disagrees with the statement that NIST was planning to cover it in the August 2010 final document. That document states that follow-on activities of the Cybersecurity Working Group need to address cyber-physical attacks.

   Therefore, we suggest changing the wording of the recommendation to "To ensure that NIST's smart grid cybersecurity guidelines will be as effective as intended, we

Mr. Gregory C. Wilshusen
Page 2

> recommend that the Director of NIST finalize the agency's plan for updating and maintaining the cybersecurity guidelines, including ensuring it incorporates (1) an assessment of the types and effects of combined cyber-physical attacks and (2) specific milestones for when efforts are to be completed. We also recommend that NIST, as a part of finalizing the plan, assess whether any cybersecurity challenges identified in this report should be addressed in the guidelines."

Please contact Rachel Kinney, NIST Management and Program Analyst, at (301) 975-8707, if you have any questions regarding this response. We look forward to receiving your final report.

Sincerely,

Gary Locke

APPENDIX V: COMMENTS FROM THE FEDERAL ENERGY REGULATORY COMMISSION

FEDERAL ENERGY REGULATORY COMMISSION
WASHINGTON, DC 20426

OFFICE OF THE CHAIRMAN

December 23, 2010

Mr. David A. Powner
Director, Information Technology Management Issues
United States Government Accountability Office
441 G Street, NW
Washington, DC 20548

Dear Mr. Powner:

Thank you for your December 1, 2010 electronic transmission of the draft report, *Electricity Grid Modernization: Progress Being Made on Cybersecurity Guidelines, but Key Challenges Remain to be Addressed*. I appreciate the opportunity to comment on this draft report.

The draft report identifies a number of challenges to securing smart grid systems and evaluates, among other things, the Federal Energy Regulatory Commission's (Commission) approach for adopting and monitoring cybersecurity and other standards for the smart grid. In general, I commend the draft report's useful discussion of cybersecurity for the electric industry, and I appreciate its helpful conclusions. I respond to the report's specific recommendations in more detail below and also present two more general issues.

In order to improve coordination among regulators and help Congress better assess the effectiveness of the process established in the Energy Independence and Security Act of 2007 (EISA) for developing voluntary smart grid standards, the draft report recommends that the Commission coordinate with state regulators and groups that represent utilities subject to limited Commission and state regulation (such as municipal and cooperative utilities) to: (1) periodically evaluate the extent to which utilities and manufacturers are following voluntary interoperability and cybersecurity standards adopted through the EISA process, and (2) develop strategies for addressing any gaps in compliance that are identified as a result of this evaluation. To the extent that the Commission determines that, despite this coordinated approach, it lacks authority to address any gaps in compliance, the draft report recommends reporting this information to Congress.

2

I agree with the recommendation to improve coordination among regulators. I will direct the Commission's staff to evaluate possible approaches for implementing this recommendation. I note that the Smart Response Collaborative of the Commission and the National Association of Regulatory Utility Commissioners recently held a technical conference on smart grid standards. I also note that the Commission must operate within its statutory authority, which may limit the tools at the Commission's disposal. Therefore, if the Commission finds that it lacks authority to address gaps in compliance with voluntary interoperability and cybersecurity standards adopted through the EISA process, I will report this information to Congress, as the draft report recommends.

The draft report also recommends that the Commission, working in conjunction with the North American Electric Reliability Corporation as appropriate, assess whether any of the challenges identified in the draft report should be addressed in Commission cybersecurity efforts. I agree with this recommendation, as well, and I have directed Commission staff to develop appropriate procedures to achieve this goal. I again note, however, that the Commission must operate within its statutory authority.

Apart from the foregoing, I would like to present two more general issues. First, the last challenge identified in the draft report is that the lack of electricity industry cybersecurity metrics makes it difficult to measure the extent to which investments in cybersecurity improve the security of smart grid systems. The draft report also recognizes experts' observation that such metrics are difficult to develop. I agree with the importance of such metrics, and would emphasize the need for the metrics to be specific and fine-tuned enough to differentiate validly among investments based on the strength of their cybersecurity protections.

Second, the draft report appears to assume that Congress intended for all relevant manufacturers and utilities to comply with the smart grid standards adopted through the EISA process. It could reasonably be argued, however, that in making these smart grid standards voluntary, Congress was not seeking to ensure this outcome. If the report is assuming a Congressional intent of uniform compliance with the standards adopted through the EISA process, it may be helpful to state more clearly both that position and the basis upon which it was reached.

3

Thank you again for the opportunity to comment on your report. Your recommendations generally represent meaningful measures to improve coordination among regulators, help Congress better assess the effectiveness of the process established in EISA for developing voluntary smart grid standards and expand the Commission cybersecurity efforts, as needed.

Sincerely,

Jon Wellinghoff
Chairman

End Notes

[1] Pub. L. No 110-140, (Dec. 19, 2007).

[2] Circuit breakers are devices used to open or close electric circuits. If a transmission or distribution line is in trouble, a circuit breaker can disconnect it from the rest of the system.

[3] GAO, Critical Infrastructure Protection: Multiple Efforts to Secure Control Systems Are Under Way, but Challenges Remain, GAO-07-1036 (Washington D.C.: Sept. 10, 2007.)

[4] We reported in 2004 that demand response—allowing customers to better understand market conditions, such as the price of electricity or limitations in supply, and respond by changing their demand for electricity—has a number of benefits. In particular, demand response programs can enhance reliability and lessen the likelihood of electricity disruptions, such as blackouts. However, our 2004 report found that, at the time, most customers lacked the necessary equipment—meters, communication devices, and special tools—for participating in demand response programs. GAO, Electricity Markets: Consumers Could Benefit from Demand Programs, but Challenges Remain, GAO-04-844. (Washington, D.C.: Aug. 13, 2004).

[5] See, for example, GAO-07-1036.

[6] GAO-07-1036.

[7] The White House, Cyberspace Policy Review: Assuring a Trusted and Resilient Information and Communications Infrastructure (Washington, D.C., May 29, 2009).

[8] FERC has the obligation to ensure that the rates charged for wholesale sales of electricity (sales of electricity for resale) by public utilities are just and reasonable and not "unduly discriminatory or preferential." See 16 U.S.C. §§ 824d, 824e. FERC is composed of up to five commissioners—including one who serves as Chairman—appointed by the President of the United States with the advice and consent of the Senate. Commissioners serve 5-year terms, and have an equal vote on regulatory matters.

[9] Prior to submission to FERC for approval, NERC standards are reviewed and voted on by members of the electricity industry who participate in NERC's FERC-approved standards development process. These standards become mandatory and enforceable in the continental United States only after they are approved by FERC. Once mandatory, both NERC and FERC have authority to enforce reliability standards.

[10] Pub. L. No. 110-140, (Dec. 19, 2007).

[11] NIST Special Publication 1108, NIST Framework and Roadmap for Smart Grid Interoperability Standards, Release 1.0, January 2010.

[12] Specifically, EISA allows FERC to adopt any standards necessary for smart grid functionality and interoperability in interstate transmission and regional and wholesale markets. According to FERC officials, if necessary, these standards may affect facilities used at the distribution level, such as smart meters, although EISA does not explicitly limit state authority over local distribution or retail sales.

[13] Pub. L. No. 111-5. (Feb. 17, 2009).

[14] In March 2009, NIST established this group, calling it the Cyber Security Coordination Task Group. In January 2010, NIST renamed it the Smart Grid Cyber Security Working Group. The working group is comprised of about 400 participants from the electricity industry, including electric companies, IT system vendors, smart grid system vendors, service providers, federal and state regulatory organizations, and academia.

[15] A coordinated cyber-physical attack involves using both cyber and physical means to attack a target. For example, a cyber attack could be aimed at disabling a security system in order to facilitate a physical attack (e.g., damaging electric grid components) against a utility's infrastructure.

[16] NIST Interagency Report 7628, Guidelines for Smart Grid Cyber Security, August 2010.

[17] Cryptography underlies many of the mechanisms used to enforce the confidentiality and integrity of critical and sensitive information. One such mechanism is encryption. Encryption can be used to provide basic confidentiality and integrity of transmitted or stored data by transforming plain text into cipher text using a special value, known as a key, and a mathematical process, known as an algorithm.

[18] Synchrophasor systems provide detailed data on the conditions of the transmission and distribution grid, which is used to improve power system reliability.

[19] NIST facilitated a process of stakeholder identification of standards to promote smart grid interoperability and cybersecurity. NIST's Cyber Security Working Group plans to evaluate whether these smart grid standards adequately address cybersecurity, including whether they align with the guidelines discussed in the previous section.

[20] Wide-area situational awareness is the visual display of broad electricity system conditions in near real time.

[21] According to FERC, its cost recovery authority for electricity investments extends to facilities used for transmission in interstate commerce. Its reliability authority applies to the bulk power system—a term that refers to facilities and control systems necessary for operating the electric transmission network and certain generation facilities needed for reliability. FERC also has regulatory authority over most of the interstate wholesale market in electricity. However, it is unclear how this authority applies to enforcement of smart grid interoperability and cybersecurity standards.

[22] Past GAO work discusses divided regulatory responsibilities in the electricity industry. GAO, Electricity Restructuring: Key Challenges Remain, GAO-06-237 (Washington, D.C.: Nov. 15, 2005), GAO, Electricity Markets: FERC's Role in Protecting Consumers. GAO-03-726R (Washington, D.C.: June 6, 2003), GAO, Electricity Restructuring: 2003 Blackout Identifies Crisis and Opportunity for the Electricity Sector, GAO-04-204 (Washington, D.C.: Nov. 18, 2003), GAO, Electricity Restructuring: Action Needed to Address Emerging Gaps in Federal Information Collection, GAO-03-586 (Washington, D.C.: June 30, 2003). Additional past GAO reports can be found at the end of this report.

[23] In one of the seven states in which we spoke with electricity regulators, the state electricity regulators did not have authority to oversee whether smart grid investments are interoperable and cyber secure. The Nebraska Power Review Board and Nebraska Public Service Commission are the primary agencies in Nebraska charged with regulating electricity. However, all utilities in Nebraska are consumer owned—such as cooperative or municipal utilities and public power districts, and neither regulator has authority to oversee whether smart grid investments are interoperable or cyber secure.

[24] GAO, Standards for Internal Control in the Federal Government, GAO/AIMD-00-21.3.1 (Washington D.C.: November 1999).

[25] GAO, Results-Oriented Government: Practices That Can Help Enhance and Sustain Collaboration among Federal Agencies GAO-06-15 (Washington D.C.: October 2005).

[26] Event logging is a capability of an IT system to record events occurring within an organization's systems and networks, including those related to computer security.

[27] The White House, Cyberspace Policy Review: Assuring a Trusted and Resilient Information and Communications Infrastructure (Washington, D.C., May 29, 2009).

[28] Metrics can be used for, among other things, measuring the effectiveness of cybersecurity controls for detecting and blocking cyber attacks.

End Notes for Appendix I

[1] NIST Interagency Report 7628, Guidelines for Smart Grid Cyber Security, August 2010.

[2] FERC, Smart Grid Policy Statement, Docket No. PL09-4-000 (Washington, D.C., July 16, 2009).

In: Cybersecurity for the Electric Smart Grid
Editors: P. Barker and R. F. Price
ISBN: 978-1-61942-353-4

Chapter 3

STATEMENT OF PATRICIA HOFFMAN ASSISTANT SECRETARY OFFICE OF ELECTRICITY DELIVERY AND ENERGY RELIABILITY U.S. DEPARTMENT OF ENERGY BEFORE THE COMMITTEE ON ENERGY AND NATURAL RESOURCES UNITED STATES SENATE*

May 5, 2011

Chairman Bingaman, Ranking Member Murkowski and members of the Committee, thank you for this opportunity to discuss the cyber security issues facing the electric industry, as well as proposed legislation intended to strengthen protection of the bulk power system and electric infrastructure from cyber security threats.

Title XIII of the Energy Independence and Security Act of 2007 (EISA) states, "It is the policy of the United States to support the modernization of the Nation's electricity transmission and distribution system to maintain a reliable and secure electricity infrastructure." The protection and resilience of critical national infrastructures is a shared responsibility of the private sector,

* This is an edited, reformatted and augmented version of statement given by Patricia HoffmanAssistant Secretary Office of Electricity Delivery and Energy Reliability U.S. Department of Energy before the Committee on Energy and Natural Resources, United States Senate, on May 5, 2011.

government, communities, and individuals. As the complexity, scale, and interconnectedness of today's infrastructures have increased, it has changed the way services and products are delivered, as well as the traditional roles of owners, operators, regulators, vendors, and customers.

Ensuring a resilient electric grid is particularly important since it is arguably the most complex and critical infrastructure that other sectors depend upon to deliver essential services. Over the past two decades, the roles of electricity sector stakeholders have shifted: generation, transmission, and delivery functions have been separated into distinct markets; customers have become generators using distributed generation technologies; and vendors have assumed new responsibilities to provide advanced technologies and improve security. These changes have created new responsibilities for all stakeholders in ensuring the continued security and resilience of the electric power grid.

CYBER SECURITY ACTIVITIES AND ACCOMPLISHMENTS

For more than a decade, the Department of Energy's Office of Electricity Delivery and Energy Reliability (OE) has been substantively engaged with the private sector to secure the electric grid. In December 2003, the Homeland Security Presidential Directive 7 (HSPD-7) designated the Department as the sector-specific agency (SSA) for the energy sector responsible for collaborating with all federal agencies, state and local governments, and the private sector. As the SSA, OE, representing the Department, works closely with the private sector and state/Federal regulators to provide secure sharing of threat information, to collaborate with industry to identify and fund gaps in infrastructure research, development and testing efforts, to conduct vulnerability assessments of the sector, and to encourage risk management strategies for critical energy infrastructure.

The 2010 *National Security Strategy* underscores the need to strengthen public-private partnerships in order to design more secure technology that will better protect and improve the resilience of critical government and industry systems and networks. OE has long recognized that neither government, nor the private sector, nor individual citizens can meet cyber security challenges alone. In 2006, OE facilitated the development of the *Roadmap to Secure Control Systems in the Energy Sector* to provide a detailed collaborative plan for improving cyber security in the energy sector and concrete steps to secure control systems used in the electricity and oil and natural gas sectors. The plan calls for a 10-year implementation timeline with a 5- year update scheduled for release in the summer of 2011. To implement the priorities in the *Roadmap*, the Energy Sector Control Systems Working Group was formed and comprised of cyber security and control systems experts from government, the electricity sector, and the oil and natural gas sector.

Since 2006, the *Roadmap* has provided a collaborative strategy for prioritizing cyber security needs and focusing actions under way throughout government and the private sector to ensure future energy system security. The *Roadmap* goals and strategy have also been fully integrated into the *Energy Sector-Specific Plan*. Since the *Roadmap* was released, important progress has been made in improving cyber security in the energy sector. These improvements have benefited existing systems and are contributing to the secure design and integration of advanced systems that incorporate smart grid technologies.

Through competitive solicitations and partnerships with industry, academia and national laboratories, OE has supported the development of several advanced cyber security technologies that are now commercially available within the energy sector:

- A technology to secure serial communications for control systems, based on the Secure Supervisory Control and Data Acquisition (SCADA) Communications Protocol developed by the Pacific Northwest National Laboratory. This technology is rapidly being adopted by utilities.
- Software toolkits, available for download from the vendor website, that let electric utilities audit the security settings of SCADA systems. The latest release addresses the Inter-Control Center Communications Protocol (ICCP), which is used for utility-to-utility communications.
- Monitoring modules that aggregate security events from a variety of data sources on the control system network and then correlate the security events to help utilities better detect cyber attacks.
- An Ethernet security gateway, based on an interoperable design developed by Sandia National Laboratories, that secures site-to-site Ethernet communications and protects private networks.

OE established the National SCADA Test Bed in 2003 to provide a national capability for cyber security experts to systematically evaluate the components of a functioning system for inherent vulnerabilities, develop mitigations, and test the effectiveness of various cyber security technologies. Major accomplishments include:

- Completed vulnerability assessments of 38 SCADA systems and provided mitigation recommendations. As a result, vendors have implemented many of the recommendations in "hardened" next-generation SCADA systems that are now commercially available and being deployed in the power grid.
- Utility groups have also formed partnerships to fund additional cyber security assessments at the test bed to address specific cyber security concerns.
- Provided advanced cyber security training for over 2300 representatives from over 200 utilities to demonstrate how to detect and respond to complex cyber attacks on SCADA systems.
- Developed the "Common Cyber Security Vulnerabilities Observed in Control System Assessments" report to help utilities and vendors mitigate vulnerabilities found in many SCADA systems. OE has also worked with the North American Electric Reliability Corporation (NERC) to develop the *Top Ten Vulnerabilities of Control Systems and their Associated Mitigations* report in 2006 and 2007.

OE is also working closely with academic and industry partners through the Trustworthy Cyber Infrastructure for the Power Grid (TCIPG), which is a University led public-private research partnership supported by OE, Department of Homeland Security (DHS), and Industry for frontier research that supports resilient and secure smart grid systems. TCIPG leverages and expands upon previous research funded primarily by the National Science Foundation. TCIPG research focuses on building trusted energy delivery control systems from un-trusted components, and transitioning next-generation cyber security technologies to the energy sector. As an example, TCIPG released the Network Access Policy Tool that is now being used by industry and asset owners to characterize the global effects of local firewall rules in control system architectures. The tool will help utilities better manage and maintain security on their highly-complex communications networks.

Just recently, OE launched several new initiatives to enhance cyber security in the energy sector.

- OE, in coordination with DHS and other Federal agencies, has conducted several cyber threat information sharing workshops to analyze classified information, determine the impact to the sector, and develop mitigations that were specifically designed to work in the sector. This cooperative process has proven to be more effective and accepted than dictating solutions to the sector.
- OE, in coordination with the National Institute of Standards and Technology (NIST) and NERC, is leading a collaborative effort with representatives from across the public and private sectors to develop a cyber security risk management guideline. The objective of this effort is to provide a consistent, repeatable, and adaptable process for the electric sector, and enable organizations to proactively manage risk.

Ensuring the cyber security of a modern, digital electricity infrastructure is a key objective of national smart grid efforts. As a result, a number of key initiatives have been developed to ensure future system security and enable the energy sector to better design, build, and integrate smart grid technologies. OE has engaged in partnerships to perform these activities with key organizations including Federal Energy Regulatory Commission (FERC), the U.S. Department of Commerce, NIST, DHS, the Federal Communications Commission, the Department of Defense (DoD), the intelligence community, the White House Office of Science and Technology Policy, state public utility commissions, the National Association of Regulatory Utility Commissioners, NERC, the Open Smart Grid Subcommittee, Electric Power Research Institute (EPRI), and other energy sector organizations.

The American Recovery and Reinvestment Act of 2009 accelerated the development of smart grid technologies by investing in pilot projects, worker training, and large scale deployments. This public-private investment worth over $9.6 billion was dedicated to a nationwide plan to modernize the electric power grid, enhance the security of U.S. energy infrastructure, and promote reliable electricity delivery. The $4.5 billion in Recovery Act funds, managed by OE, was leveraged by $5.1 billion in funds from the private sector to support 132 Smart Grid Investment Grant and Smart Grid Demonstration Grant projects across the country. Each project awardee committed to implementing a cyber security plan that includes an evaluation of cyber risks and planned mitigations, cyber security criteria for device and vendor selection, and relevant standards or best practices the project will follow.

As called for in Section 1305 of EISA, OE is collaborating with NIST and other agencies and organizations to develop a framework and roadmap for interoperability standards that includes cyber security as a critical element. As part of this effort, NIST established the public-private Smart Grid Interoperability Panel, and within that, the 450-member Cyber Security Working Group (CSWG) to lead the development of cyber security requirements for the smart grid. After engaging members in numerous workshops and teleconferences and following two formal reviews, the CSWG released the first version of its "Cyber Security Guidelines for the Smart Grid". The three-volume document details a strategy that includes smart grid use cases, a high-level smart grid risk assessment process, smart grid-specific security requirements, development of a security architecture, assessment of smart grid standards, and development of a conformity assessment program for requirements.

To address cyber security needs for smart grid technologies, OE partnered with leading utilities and EPRI to develop cyber security profiles for major smart grid applications – Advanced Metering Infrastructure, Third-Party Data Access, and Distribution Automation. These profiles provide vendor-neutral, actionable guidance to utilities, vendors and government entities on how to build cyber security into smart grid components in the development stage, and how to implement those safeguards when the components are integrated into the power grid. These documents support the NIST "Cyber Security Guidelines for the Smart Grid" NISTIR – 7628. OE also co-chairs the NIST CSWG.

Senate Energy and Natural Resources Committee Proposed Legislation

The proposed bill includes provisions intended to strengthen the bulk power system and electric infrastructure by addressing cyber security vulnerabilities and protecting against cyber security threats by adding a new section to the Federal Power Act (FPA). While the Administration does not yet have a position on the bill, the Department offers the following observations.

To begin with, the proposed bill correctly identifies, defines, and distinguishes between a cyber security vulnerability and a cyber security threat. These are two related, but different concepts. Vulnerabilities need to be identified and addressed, while threats need to be protected against. In that regard, references in the proposed bill to "protecting critical electric infrastructure from cyber security vulnerabilities" should be changed to "addressing critical electric infrastructure cyber security vulnerabilities."

In addition, Section 224(a)(1) defines critical electric infrastructure to include distribution assets that affect interstate commerce. This significantly expands FERC's jurisdiction for setting reliability standards beyond the bulk power system as provided in FPA section 215.

Also, Section 224(f) would require a comprehensive plan identifying emergency measures to protect the reliability of the electric power supply of national defense facilities located in Alaska, Hawaii, and Guam in the event of an imminent cyber security threat. Pertinent to that, in July 2010, DOE and DoD signed a memorandum of understanding (MOU) "Concerning Cooperation in a Strategic Partnership to Enhance Energy Security". The purpose of the MOU is to enhance national energy security and demonstrate Federal Government leadership in transitioning America to a low carbon economy. This MOU provides an opportunity to develop a comprehensive approach that reduces the impact of power loss to defense critical assets, considering both mitigation and response measures to ensure vital defense capabilities are not disrupted.

Finally, the legislation does not yet address a unique, sensitive cyber security information disclosure problem faced by Federal Power Marketing Administrations subject to both the Freedom of Information Act and mandatory reliability standards enacted under Section 215 of the Federal Power Act. This sensitive information, developed under the mandatory reliability standards, appears not to be protected from public disclosure under the Freedom of Information Act. This security vulnerability could be avoided if legislation providing statutory protection for this information were enacted that qualified under Exemption 3 of the Freedom of Information Act.

CONCLUSION

In conclusion, I would like to again thank this Committee for its leadership in supporting the protection of the bulk power system and critical electric infrastructure against cyber security threats. Recognizing the interdependencies between different sectors, it is important to have a comprehensive strategy for cyber security legislation. DOE would be happy to work with the Committee on this legislation.

I would be pleased to address any questions the Committee might have.

In: Cybersecurity for the Electric Smart Grid ISBN: 978-1-61942-353-4
Editors: P. Barker and R. F. Price

Chapter 4

TESTIMONY OF JOSEPH MCCLELL AND DIRECTOR, OFFICE OF ELECTRIC RELIABILITY FEDERAL ENERGY REGULATORY COMMISSION BEFORE THE COMMITTEE ON ENERGY AND NATURAL RESOURCES UNITED STATES SENATE*

May 5, 2011

Mr. Chairman and Members of the Committee:

Thank you for this opportunity to appear before you to discuss the security of the electric grid. My name is Joseph McClelland. I am the Director of the Office of Electric Reliability (OER) of the Federal Energy Regulatory Commission (FERC or Commission). The Commission's role with respect to reliability is to help protect and improve the reliability of the Nation's bulk power system through effective regulatory oversight as established in the Energy Policy Act of 2005. I am here today as a Commission staff witness and my remarks do not necessarily represent the views of the Commission or any individual Commissioner.

* This is an edited, reformatted and augmented version of testimony given by Joseph Mcclelland Director, Office of Electric Reliability Federal Energy Regulatory Commission before the Committee on Energy and Natural Resources, United States Senate, on May 5, 2011.

My testimony summarizes the Commission's oversight of the reliability of the electric grid under section 215 of the Federal Power Act (FPA) and the Commission's implementation of that authority with respect to cyber security primarily through Order No. 706. I also will describe some of the current limitations in Federal authority to protect the grid against physical and cyber security threats, and also comment on the cyber security discussion draft. The Commission currently does not have sufficient authority to require effective protection of the grid against cyber or physical attacks. If adequate protection is to be provided, legislation is needed and my testimony discusses the key elements that should be included in legislation in this area.

BACKGROUND

In the Energy Policy Act of 2005 (EPAct 2005), Congress entrusted the Commission with a major new responsibility to oversee mandatory, enforceable reliability standards for the Nation's bulk power system (excluding Alaska and Hawaii). This authority is in section 215 of the Federal Power Act. Section 215 requires the Commission to select an Electric Reliability Organization (ERO) that is responsible for proposing, for Commission review and approval, reliability standards or modifications to existing reliability standards to help protect and improve the reliability of the Nation's bulk power system. The Commission has certified the North American Electric Reliability Corporation (NERC) as the ERO. The reliability standards apply to the users, owners and operators of the bulk power system and become mandatory in the United States only after Commission approval. The ERO also is authorized to impose, after notice and opportunity for a hearing, penalties for violations of the reliability standards, subject to Commission review and approval. The ERO may delegate certain responsibilities to "Regional Entities," subject to Commission approval.

The Commission may approve proposed reliability standards or modifications to previously approved standards if it finds them "just, reasonable, not unduly discriminatory or preferential, and in the public interest." The Commission itself does not have authority to modify proposed standards. Rather, if the Commission disapproves a proposed standard or modification, section 215 requires the Commission to remand it to the ERO for further consideration. The Commission, upon its own motion or upon complaint, may direct the ERO to submit a proposed standard or modification on a specific matter but it does not have the authority to modify or author a standard and must depend upon the ERO to do so.

Limitations of Section 215 and the Term "Bulk Power System"

Currently, the Commission's jurisdiction and reliability authority is limited to the "bulk power system," as defined in the FPA, and therefore excludes Alaska and Hawaii, including any federal installations located therein. The current interpretation of "bulk power system" also excludes some transmission and all local distribution facilities, including virtually all of the grid facilities in certain large cities such as New York, thus precluding Commission action to mitigate cyber or other national security threats to reliability that involve such facilities and major population areas. The Commission recently issued Order No. 743, which directs NERC to revise its interpretation of the bulk power system to eliminate inconsistencies across regions, eliminate the ambiguity created by the current discretion in NERC's definition of bulk electric system, provide a backstop review to ensure that any variations do not compromise reliability, and ensure that facilities that could significantly affect reliability are subject to mandatory rules. NERC is currently developing its response to that order. However, it is important to note that section 215 of the FPA excludes local distribution facilities from the Commission's reliability jurisdiction, so any revised bulk electric system definition developed by NERC will still not apply to local distribution facilities.

Critical Infrastructure Protection Reliability Standards

An important part of the Commission's current responsibility to oversee the development of reliability standards for the bulk power system involves cyber security. In August 2006, NERC submitted eight proposed cyber security standards, known as the Critical Infrastructure Protection (CIP) standards, to the Commission for approval under section 215. Critical infrastructure, as defined by NERC for purposes of the CIP standards, includes facilities, systems, and equipment which, if destroyed, degraded, or otherwise rendered unavailable, would affect the reliability or operability of the "Bulk Electric System." Under NERC's implementation plan for the CIP standards, full compliance became mandatory on July 1, 2010.

On January 18, 2008, the Commission issued Order No. 706, the Final Rule approving the CIP reliability standards while concurrently directing NERC to develop significant modifications addressing specific concerns. The Commission set a deadline of July 1, 2009 for NERC to resolve certain issues in the CIP reliability standards, including deletion of the "reasonable business judgment" and "acceptance of risk" language in each of the standards. NERC concluded that this deadline would create a very compressed schedule for its stakeholder process. Therefore, it divided all of the changes directed by the Commission into phases, based on their complexity. NERC opted to resolve the simplest changes in the first phase, while putting off more complex changes for later versions.

NERC filed the first phase of the modifications to the CIP Reliability Standards (Version 2) on May 22, 2009. In this phase, NERC removed from the standards the terms "reasonable business judgment" and "acceptance of risk," added a requirement for a "single senior manager" responsible for CIP compliance, and made certain other administrative and clarifying changes. In a September 30, 2009 order, the Commission approved the Version 2 CIP standards and directed NERC to develop additional modifications to certain of them. Pursuant to the Commission's September 30, 2009 order, NERC submitted Version 3 of the CIP standards which revised Version 2 as directed. The Version 3 CIP standards became effective on October 1, 2010. This first phase of the modifications directed by the Commission in Order No. 706, which encompassed both Version 2 and Version 3, did not modify the critical asset identification process, a central concern in Order No. 706.

On February 10, 2011, NERC initiated the second phase of the Order No. 706 directed modification, filing a petition seeking approval of Version 4 of the CIP standards. Version 4 includes new proposed criteria to identify "critical assets" for purposes of the CIP reliability standards. This filing is currently under review by the Commission. In order to better understand the NERC Version 4 petition, particularly the number of critical cyber assets that will be identified under this revision, the Commission issued data requests to NERC, with responses due on July 11, 2011, which reflects an extension of time requested by NERC.

The remaining CIP standards revisions to respond to the Commission's directives issued in Order No. 706 are still under development by NERC. It is important to note that the majority of the Order No. 706 directed modifications to the CIP standards have yet to be addressed by NERC. Until they are addressed, there are significant gaps in protection such as a needed requirement for a defense in depth posture. NERC's standards development plan filed with the Commission in April 2011 classifies these outstanding revisions to the CIP standards as "High Priority" with a targeted completion in the second quarter of 2012.

Identification of Critical Assets

As currently written, the CIP reliability standards allow utilities significant discretion to determine which of their facilities are "critical assets and the associated critical cyber assets," and therefore are subject to the requirements of the standards. In Order No. 706, the Commission directed NERC to revise the standards to require independent oversight of a utility's decisions by industry entities with a "wide-area view," such as reliability coordinators or the Regional Entities, subject to the review of the Commission. This revision to the standards, like all revisions, is subject to approval by the affected stakeholders in the standards development process. NERC has attempted to address this directive in Version 4 of the CIP standards, which is now under review by the Commission.

When, in Order No. 706, the Commission approved Version 1 of the CIP reliability standards, it also required entities under those standards to self-certify their compliance progress every six months. In December 2008, NERC conducted a self-certification study, asking each entity to report limited information on its critical assets and the associated critical cyber assets identified in compliance with reliability standard CIP-002-1. As the Commission stated in Order No. 706, the identification of critical assets is the cornerstone of the CIP standards. If that identification is not done well, the CIP standards will be ineffective at protecting the bulk power system. The results of NERC's self-certification request showed that only 29% of responding generation owners and operators identified at least one critical asset, while about 63% of the responding transmission owners identified at least one critical asset. NERC expressed its concern with these results in a letter to industry stakeholders dated April 7, 2009.

NERC conducted another self-certification survey of responsible entities to determine progress towards identification of critical cyber assets. It gathered information about critical assets and critical cyber assets as of December 31, 2009. This survey included additional questions designed to obtain a better understanding of the results from industry's critical asset identification process. In general, this survey did not demonstrate a significant increase in identified critical assets. NERC noted some encouraging results as well as some that were a cause for concern. In addition, the Regional Entities have been performing audits which have included registered entities' determination of their critical cyber asset lists. FERC staff has been observing selected audits to examine the Regional Entities' methods of conducting these audits. It is important to note that although "critical assets" are used to identify subsequent "critical cyber assets," only the subset of "critical cyber assets" are subject to the CIP standards.

NERC's Critical Infrastructure Protection Committee released a guidance document to assist registered entities in identifying their critical assets. That document, which took effect on September 17, 2009, provides "guidelines" that define which assets should be evaluated, provides risk-based evaluation guidance for determining critical assets, and describes reasonable bases that could be used to support that determination. A second NERC security guideline regarding critical cyber assets became effective on June 17, 2010. This security guideline "provides guidance for identifying Critical Cyber Assets by evaluating potential impacts to 'reliable operation' of a Critical Asset." Neither of these guidance documents contained any actions that were mandatory for users, owners or operators of the bulk-power system.

Version 4 of the CIP standards, which are currently pending before the Commission, would change the way in which critical assets are identified. Instead of using a loosely defined risk-based assessment methodology, CIP-002 Version 4 Attachment 1 contains what NERC describes as "uniform criteria for the identification of Critical Assets." For example, criterion 1.1 would identify generation plants equal to or greater than 1500MW as critical assets. The filing asserts that this would account for 29% of the installed generator capacity in the United States. Because this is an on-going proceeding before the Commission, I am limited in what I can discuss about the merits of NERC's petition.

The NERC Process

As an initial matter, it is important to recognize how mandatory reliability standards are established. Under section 215, reliability standards must be developed by the ERO through an open, inclusive, and public process. The Commission can direct NERC to develop a reliability standard to address a particular reliability matter, including cyber security threats or vulnerabilities.

However, the NERC process typically requires years to develop standards for the Commission's review. In fact, the CIP standards approved by the Commission in January 2008 took approximately three years to develop.

NERC's procedures for developing standards allow extensive opportunity for stakeholder comment, are open, and are generally based on the procedures of the American National Standards Institute. The NERC process is intended to develop consensus on both the need for, and the substance of, the proposed standard. Although inclusive, the process is relatively slow, open and unpredictable in its responsiveness to the Commission's directives. This process requires public disclosure regarding the reason for the proposed standard, the manner in which the standard will address the issues, and any subsequent comments and resulting modifications in the standards as the affected stakeholders review the material and provide comments. NERC-approved standards are then submitted to the Commission for its review.

The procedures used by NERC are appropriate for developing and approving routine reliability standards. The process allows extensive opportunities for industry and public comment. The public nature of the reliability standards development process can be a strength of the process. However, it can be an impediment when measures or actions need to be taken to address threats to national security quickly, effectively and in a manner that protects against the disclosure of security-sensitive information. The current procedures used under section 215 for the development and approval of reliability standards do not provide an effective and timely means of addressing urgent cyber or other national security risks to the bulk power system, particularly in emergency situations. Certain circumstances, such as those involving national security, may require immediate action, while the reliability standard procedures take too long to implement efficient and timely corrective steps. On September 3, 2010, FERC approved a new reliability standards process manual filed by NERC. While this manual includes a process for developing a standard related to a confidential issue, the new process is untested and it is unclear how the process would be implemented.

FERC rules governing review and establishment of reliability standards allow the agency to direct the ERO to develop and propose reliability standards under an expedited schedule. For example, FERC could order the ERO to submit a reliability standard to address a reliability vulnerability within 60 days. Also, NERC's rules of procedure include a provision for approval of "urgent action" standards that can be completed within 60 days and which may be further expedited by a written finding by the NERC board of trustees that an extraordinary and immediate threat exists to bulk power system reliability or national security. However, it is not clear NERC could meet this schedule in practice. Moreover, faced with a national security threat to reliability, there may be a need to act decisively in hours or days, rather than weeks, months or years. That would not be feasible even under the urgent action process. In the meantime, the bulk power system would be left vulnerable to a known national security threat. Moreover, existing procedures, including the urgent action procedure, could widely publicize both the vulnerability and the proposed solutions, thus increasing the risk of hostile actions before the appropriate solutions are implemented.

In addition, a reliability standard submitted to the Commission by NERC may not be sufficient to address the identified vulnerability or threat. Since FERC may not directly modify a proposed reliability standard under section 215 and must either approve or remand it, FERC would have the choice of approving an inadequate standard and directing changes, which reinitiates a process that can take years, or rejecting the standard altogether. Under either approach, the bulk power system would remain vulnerable for a prolonged period.

This concern was highlighted in the Department of Energy Inspector General's January 2011 audit report on FERC's "Monitoring of Power Grid Cyber Security." The audit report identified concerns regarding the adequacy of the CIP standards and the implementation and schedule for the CIP standards, and concluded that these problems exist, in part, because the Commission's authority to ensure adequate cyber security over the bulk electric system is limited. The audit report concludes that the Commission should take a more aggressive action when ordering new or revised standards and highlights its lack of authority to implement its own reliability standards or mandatory alerts in response to emerging threats or vulnerabilities. This report emphasizes the need for FERC to have additional authority for ensuring adequate cyber security over the bulk electric system.

Finally, the open and inclusive process required for standards development is not consistent with the need to protect security-sensitive information. For instance, a formal request for a new standard would normally detail the need for the standard as well as the proposed mitigation to address the issue, and the NERC-approved version of the standard would be filed with the Commission for review. This public information could help potential adversaries in planning attacks.

NERC's Formal Notices

Currently, the alternative to a mandatory reliability standard is for NERC to issue a formal notice encouraging utilities and others to take voluntary action to guard against a specific cyber or other vulnerability. Such a notice may be an Advisory, a Recommendation or an Essential Action. The notice approach allows for quicker action, but compliance with a notice is voluntary, and will likely produce inconsistent and potentially ineffective responses. For example, two Advisories and a Recommendation were issued in 2010 by NERC, regarding an identified cyber security threat referred to as "Stuxnet." The details of actions taken to mitigate the vulnerabilities identified by Stuxnet, and the assets to which they apply, as well as their effectiveness, are not known. Reliance on voluntary measures to protect national security is fundamentally inconsistent with the conclusion Congress reached during enactment of EPAct 2005, that voluntary standards are not sufficient to protect the reliability of the bulk power system.

Smart Grid

The need for vigilance will increase as new technologies are added to the bulk power system. For example, smart grid technology promises significant benefits in the use of electricity. These include the ability to better manage not only energy sources but also energy consumption. However, a smarter grid would permit two-way communication between the electric system and a large number of devices located outside of controlled utility environments, which will introduce many potential access points.

Smart grid applications will automate many decisions on the supply and use of electricity to increase efficiencies and ultimately to allow cost savings. Without adequate physical and cyber protections, however, this level of automation may allow adversaries to gain access to the rest of the company's data and control systems and cause significant harm. Security features must be an integral consideration when developing smart grid technology and must be assured before widespread installation of new equipment. The challenge will be to focus not only on general approaches but, importantly, on the details of specific technologies and the risks they may present.

Regarding data, there are multiple ways in which smart grid technologies may introduce new cyber vulnerabilities into the system. For example an attacker could gain access to a remote or intermediate smart grid device and change data values monitored or received from downstream devices, and pass the incorrect data up-stream to cause operators or automatic programs to take incorrect actions.

In regard to control systems, an attacker that gains access to the communication channels could order metering devices to disconnect customers, order previously shed load to come back on line prematurely, or order dispersed generation sources to turn off during periods when load is approaching generation capacity, causing instability and outages on the bulk power system. One of the potential capabilities of the smart grid is the ability to remotely disconnect service using advanced metering infrastructure (AMI). If insufficient security measures are implemented in a company's AMI application, an adversary may be able to access the AMI system and could conceivably disconnect every customer with an AMI device. If such an attack is widespread enough, the resultant disconnection of load on the distribution system could result in impacts to the bulk power system. If an adversary follows this disconnection event with a subsequent and targeted cyber attack against remote meters, the restoration of service could be greatly delayed.

In addition to any smart grid related standards that may be adopted by the Commission, the CIP standards will apply to some, but not most, smart grid applications. The standards require users, owners and operators of the bulk power system to protect cyber assets, including hardware, software and data, which would affect the reliability or operability of the bulk power system. These assets are identified using a risk-based assessment methodology that identifies electric assets that are critical to the reliable operation of the bulk power system. If a smart grid device were to control a critical part of the bulk power system, it should be considered a critical cyber asset subject to the protection requirements of the CIP standards. However, this designation is currently up to the affected entity as part of its self-determination of critical cyber assets, as discussed previously.

Many of the smart grid applications will be deployed at the distribution and end-user level. For example, some applications may be targeted at improving market efficiency in ways that may not have a reliability impact on the bulk power system, such that the protection requirements of the CIP standards, as they are currently written, may not apply. However, as discussed above, these applications either individually or in the aggregate could affect the bulk power system.

PHYSICAL SECURITY AND OTHER THREATS TO RELIABILITY

The existing reliability standards do not extend to physical threats to the grid, but physical threats can cause equal or greater destruction than cyber attacks and the Federal government should have no less ability to act to protect against such potential damage. One example of a physical threat is an electromagnetic pulse (EMP) event. In 2001, Congress established a commission to assess the threat from EMP, with particular attention to be paid to the nature and magnitude of high-altitude EMP threats to the United States; vulnerabilities of U.S. military and civilian infrastructure to such attack; capabilities to recover from an attack; and the feasibility and cost of protecting military and civilian infrastructure, including energy infrastructure. In 2004, the EMP commission issued a report describing the nature of EMP attacks, vulnerabilities to EMP attacks, and strategies to respond to an attack.[1] A second report was produced in 2008 that further investigated vulnerabilities of the Nation's infrastructure to EMP.[2] Both electrical equipment and control systems can be damaged by EMP.

An EMP may also be a naturally-occurring event caused by solar flares and storms disrupting the Earth's magnetic field. In 1859, a major solar storm occurred, causing auroral displays and significant shifts of the Earth's magnetic fields. As a result, telegraphs were rendered useless and several telegraph stations burned down. The impacts of that storm were muted because semiconductor technology did not exist at the time. Were the storm to happen today, according to an article in *Scientific American*, it could "severely damage satellites, disable radio communications, and cause continent-wide electrical black-outs that would require weeks or longer to recover from."[3] Although storms of this magnitude occur rarely, storms and flares of lesser intensity occur more frequently. Storms of about half the intensity of the 1859 storm occur every 50 years or so according to the authors of the *Scientific American* article, and the last such storm occurred in November 1960, leading to world-wide geomagnetic disturbances and radio outages. The power grid is particularly vulnerable to solar storms, as transformers are electrically grounded to the Earth and susceptible to damage from geomagnetically induced currents. The damage or destruction of numerous transformers across the country would result in reduced grid functionality and even prolonged power outages.

In March 2010, Oak Ridge National Laboratory (Oak Ridge) and their subcontractor Metatech released a study that explored the vulnerability of the electric grid to EMP-related events. This study was a joint effort contracted by FERC staff, the Department of Energy and the Department of Homeland Security and expanded on the information developed in other initiatives, including the EMP commission reports. The series of reports provided detailed technical background and outlined which sections of the power grid are most vulnerable, what equipment would be affected, and what damage could result. Protection concepts for each threat and additional methods for remediation were also included along with suggestions for mitigation. The results of the study support the general conclusion that EMP events pose substantial risk to equipment and operation of the Nation's power grid and under extreme conditions could result in major long term electrical outages. In fact, solar magnetic disturbances are inevitable with only the timing and magnitude subject to variability. The study assessed the 1921 solar storm, which has been termed a 1-in-100 year event, and applied it to today's power grid. The study concluded that such a storm could damage or destroy up to 300 bulk power system transformers interrupting service to 130 million people for a period of years.

The existing reliability standards do not address EMP vulnerabilities. Protecting the electric generation, transmission and distribution systems from severe damage due to an EMPrelated event would involve vulnerability assessments at every level of electric infrastructure.

The Need for Legislation

In my view, section 215 of the Federal Power Act provides an adequate statutory foundation for the ERO to develop most reliability standards for the bulk power system. However, the nature of a national security threat by entities intent on attacking the U.S. through vulnerabilities in its electric grid stands in stark contrast to other major reliability vulnerabilities that have caused regional blackouts and reliability failures in the past, such as vegetation management and protective relay maintenance practices. Widespread disruption of electric service can quickly undermine the U.S. government, its military, and the economy, as well as endanger the health and safety of millions of citizens. Given the national security dimension to this threat, there may be a need to act quickly to protect the grid, to act in a manner where action is mandatory rather than voluntary, and to protect certain information from public disclosure.

The Commission's current legal authority is inadequate for such action. This is true of both cyber and physical threats to the bulk power system that pose national security concerns.

Any new legislation should address several key concerns. First, to prevent a significant risk of disruption to the grid, legislation should allow the Commission to take action before a cyber or physical national security incident has occurred. In my opinion, the cyber security discussion draft addresses this concern by allowing the Commission to timely act on cyber security vulnerabilities before an incident occurs and by giving the Secretary of Energy emergency authority to act on cyber security threats. In particular, the Commission should be able to require mitigation even before or while NERC and its stakeholders develop a standard, when circumstances require urgent action.

Second, any legislation should allow the Commission to maintain appropriate confidentiality of sensitive information submitted, developed or issued under this authority. Without such confidentiality, the grid may be more vulnerable to attack and the Commission will not be able to adequately protect it. The cyber security discussion draft also includes provisions for protection of critical electric infrastructure information, which includes a provision for FERC to establish procedures to allow the Commission to release critical infrastructure information to the extent necessary to enable entities to implement any FERC order under the proposal. It also appropriately would require FERC to limit redistribution of information so that the information is only in the hands of those that need to know.

Third, if additional reliability authority is limited to the bulk power system, as that term is currently defined in the FPA, it would not authorize Commission action to mitigate cyber or other national security threats to reliability that involve certain critical facilities and major population areas. The cyber security discussion draft would apply to any entity that owns, controls, or operates critical electric infrastructure. While Alaska and Hawaii would be excluded, the discussion draft requires the Secretary of Defense to prepare a comprehensive plan to protect any national defense facilities located in those states.

Fourth, it is important that entities be able to recover costs they incur to mitigate vulnerabilities and threats. The cyber security discussion draft requires the Commission to permit public utilities to recover prudently incurred costs required to implement immediate actions ordered by the Secretary of Energy to avert or mitigate a cyber security threat. I support this provision and any clarifications that might better ensure recovery of costs incurred under this legislation.

Finally, in my view, any legislation on national security threats to reliability should address not only cyber security threats but also natural events; i.e., a geomagnetic disturbance, or intentional physical malicious acts (targeting, for example, critical substations and generating stations) including threats from an electromagnetic pulse. This additional authority would not displace other means of protecting the grid, such as action by federal, state and local law enforcement and the National Guard. If particular circumstances cause both FERC and other governmental authorities to require action by utilities, FERC would coordinate with other authorities as appropriate.

In short, any new authority should allow the Commission to quickly order mandatory measures that are focused and confidential to address fast-moving, sophisticated and targeted cyber and physical attacks and natural events while providing cost recovery to the affected entities.

CONCLUSION

The Commission's current authority is not adequate to address cyber or other national security threats to the reliability of our transmission and power system. These types of threats pose an increasing risk to our Nation's electric grid, which undergirds our government and economy and helps ensure the health and welfare of our citizens. Congress should address this risk now. The cyber security discussion draft in front of us today would go a long way to resolving this issue. Thank you again for the opportunity to testify today. I would be happy to answer any questions you may have.

End Notes

[1] Graham, Dr. William R. et al., *Report of the Commission to Assess the Threat to the United States from Electromagnetic Pulse (EMP) Attack* (2004).

[2] Dr. John S., Jr. et al., *Report of the Commission to Assess the Threat to the United States from Electromagnetic Pulse (EMP) Attack* (2008).

[3] Odenwald, Sten F. and Green, James L., *Bracing the Satellite Infrastructure for a Solar Superstorm*, Scientific American Magazine (Jul. 28, 2008).

In: Cybersecurity for the Electric Smart Grid ISBN: 978-1-61942-353-4
Editors: P. Barker and R. F. Price

Chapter 5

TESTIMONY OF GERRY CAULEY, PRESIDENT AND CHIEF EXECUTIVE OFFICER, NORTH AMERICAN ELECTRIC RELIABILITY CORPORATION SENATE ENERGY AND NATURAL RESOURCES COMMITTEE HEARING TO RECEIVE TESTIMONY ON A JOINT STAFF DISCUSSION DRAFT PERTAINING TO CYBERSECURITY OF THE BULK-POWER SYSTEM AND ELECTRIC INFRASTRUCTURE*

May 5, 2011

INTRODUCTION

Good morning Chairman Bingaman, Ranking Member Murkowski, members of the Committee and fellow panelists. My name is Gerry Cauley and I am the President and CEO of the North American Electric Reliability

* This is an edited, reformatted and augmented version of testimony given by Gerry Cauley, President and Chief Executive Officer, North American Electric Reliability Corporation at Senate Energy and Natural Resources Committee Hearing to Receive Testimony on a Joint Staff Discussion Draft Pertaining to Cybersecurity of the Bulk-Power System and Electric Infrastructure, on May 5, 2011.

Corporation (NERC). I am a graduate of the U.S. Military Academy, a former officer in the U.S. Army Corps of Engineers, and have more than 30 years' experience in the bulk power system[1] industry, including service as a lead investigator of the August 2003 Northeast blackout and coordinator of the NERC Y2K program. I appreciate the opportunity to testify today on the discussion draft of cybersecurity legislation.

NERC's Mission

NERC's mission is to ensure the reliability of the bulk power system of North America and promote reliability excellence. NERC was founded in 1968 to develop voluntary standards for the owners and operators of the bulk power system. NERC is an independent corporation whose membership includes large and small electricity consumers, government representatives, municipalities, cooperatives, independent power producers, investor-owned utilities, independent transmission system operators and federal power marketing agencies such as TVA and Bonneville Power Administration.

In 2007, NERC was designated the Electric Reliability Organization (ERO) by the Federal Energy Regulatory Commission (FERC) in accordance with Section 215 of the Federal Power Act (FPA), enacted by the Energy Policy Act of 2005. Upon approval by FERC, NERC's reliability standards became mandatory within the United States. These mandatory reliability standards include Critical Infrastructure Protection (CIP) Standards 001 through 009, which address the security of cyber assets essential to the reliable operation of the electric grid. To date, these standards (and those promulgated by the Nuclear Regulatory Commission) are the only mandatory cybersecurity standards in place across the critical infrastructures of the United States. Subject to FERC oversight, NERC and its Regional Entity partners enforce these standards, which are developed with substantial input from industry and approved by FERC, to accomplish our mission to ensure the reliability of the electric grid. In its position between industry and government, NERC embodies the often-invoked goal of creating effective partnerships between the public sector and the private sector.

As a result of society's growing dependence on electricity, the electric grid is one of the Nation's most critical infrastructures. The bulk power system in North America is one of the largest, most complex, and most robust systems ever created by mankind. Throughout North America, four interconnections with a capacity of over one-million megawatts of generation and nearly half-a-million miles of high voltage transmission lines all acting in unison, meet the electric needs of more than 340 million people, with a maximum demand of nearly 850 thousand megawatts. The electricity being used in this room right now is generated and transmitted in real time over a complex series of lines and stations from as far away as Ontario or Tennessee. As complex as it is, few machines are as robust as the bulk power system. Decades of experience with hurricanes, ice storms and other natural disasters, as well as mechanical breakdowns, vandalism and sabotage, have taught the electric industry how to build strong and reliable networks that generally withstand all but the worst natural and physical disasters while supporting affordable electric service. The knowledge that disturbances on the grid can impact operations thousands of miles away has influenced the electric industry culture of reliability, affecting how it plans, operates and protects the bulk power system.

The Cybersecurity Challenge for the Grid and NERC's Approach to Addressing It

Along with the rest of our economy, the electric industry has become increasingly dependent on digital technology to reduce costs, increase efficiency and maintain the reliability of the bulk power system. The networks and computer environments that make up this digital technology could be as vulnerable to malicious attacks and misuse as any other technology infrastructure. Much like the defense of this country, the defense of the bulk power system requires constant vigilance and expertise.

As CEO of the organization charged with overseeing the reliability and security of the North American grid, I am deeply concerned about the changing risk landscape from conventional risks, such as extreme weather and equipment failures, to new and emerging risks where we are left to imagine scenarios that might occur and prepare to avoid or mitigate the consequences. Some of those consequences could be much more severe than we have previously experienced. I am most concerned about coordinated physical and cyber attacks intended to disable elements of the power grid or deny electricity to specific targets, such as government or business centers, military installations, or other infrastructures. These threats differ from conventional risks in that they result from intentional actions by adversaries and are not simply random failures or acts of nature.

The most effective approach against such adversaries is through thoughtful application of resiliency principles, as outlined in a National

Infrastructure Advisory Council (NIAC) report on the grid delivered to the White House in October 2010. I served on that council along with a number of industry CEOs. Resiliency requires proactive readiness for whatever may come our way and includes robustness; the ability to minimize consequences in real-time; the ability to restore essential services; and the ability to adapt and learn. Examples of the NIAC team's recommendations include: 1) a national response plan that clarifies the roles and responsibilities between industry and government; 2) improved sharing of actionable information by government regarding threats and vulnerabilities; 3) cost recovery for security investments driven by national policy; and 4) a strategy on spare equipment with long lead times, such as electric power transformers.

Critical Infrastructure Protection ("CIP") Reliability Standards and other NERC Measures to Address Cybersecurity Threats and Vulnerabilities

NERC's critical infrastructure program, including both reliability standards and alerts, provides many tools to respond to cyber threats and vulnerabilities. Industry, consumers, and government representatives all participate in the NERC standards development process and provide important expertise.

1. Reliability Standards

NERC has nine existing CIP standards that address the following areas:

- Standard CIP-001: Covers Sabotage Reporting.
- Standard CIP-002: Requires the identification and documentation of the Critical Cyber Assets associated with the Critical Assets that support the reliable operation of the Bulk Electric System.
- Standard CIP-003: Requires that Responsible Entities have minimum security management controls in place to protect Critical Cyber Assets.
- Standard CIP-004: Requires that personnel having authorized cyber or authorized unescorted physical access to Critical Cyber Assets, including contractors and service vendors, have an appropriate level of personnel risk assessment, training, and security awareness.

- Standard CIP-005: Requires the identification and protection of the Electronic Security Perimeter(s) inside which all Critical Cyber Assets reside, as well as all access points on the perimeter.
- Standard CIP-006: Intended to ensure the implementation of a physical security program for the protection of Critical Cyber Assets.
- Standard CIP-007: Requires Responsible Entities to define methods, processes, and procedures for securing those systems determined to be Critical Cyber Assets, as well as the other (non-critical) Cyber Assets within the Electronic Security Perimeter(s).
- Standard CIP-008: Ensures the identification, classification, response, and reporting of Cyber Security Incidents related to Critical Cyber Assets.
- Standard CIP-009: Ensures that recovery plan(s) are put in place for Critical Cyber Assets and that these plans follow established business continuity and disaster recovery techniques and practices.

In December 2010, NERC approved an enhancement to its Critical Cyber Asset Identification standard (CIP-002 version 4) that establishes bright-line criteria for the identification of critical assets. This enhanced standard was filed with FERC in February 2011 and is currently pending FERC approval.

In addition to the development of reliability standards through NERC's regular processes, FERC has authorized NERC to use an expedited standards development process to meet urgent reliability issues. NERC also has rules approved by FERC to enable the development of special standards on an expedited, confidential basis to address imminent or longer term national security threats.

Finally, FERC can order NERC to develop a proposed reliability standard or a modification to a reliability standard to address a specific matter (such as a cyber threat or vulnerability) under FPA Section 215(d)(5). In addition, the NERC Board of Trustees may propose and adopt a standard in response to a FERC directive if the board determines that the regular standards process is not being sufficiently responsive to the Commission.

Compliance with the NERC CIP standards is an important threshold for properly securing the BPS. However, there is no single security asset, security technique, security procedure or security standard that, even if strictly followed or complied with, will protect an entity from all potential threats. The cybersecurity threat environment is constantly changing and our defenses must keep pace. Security best-practices call for additional processes, procedures and technologies beyond those required by the CIP standards.

2. NERC Alerts

Not all vulnerabilities can or should be addressed through a reliability standard. In such cases, NERC Alerts are a key element in critical infrastructure protection. To address cyber challenges not covered under the CIP Standards, NERC works through its Electricity Sector-Information Sharing and Analysis Center (ES-ISAC) to inform the industry and recommend preventative actions.

NERC must be able to promptly disseminate threat indications, analyses and warnings to assist electricity-sector participants in taking protective actions. NERC staff with appropriate security clearances often work with cleared personnel from Federal agencies to communicate sanitized sensitive information to the industry. As defined in NERC's Rules of Procedure, the ES-ISAC developed the following three levels of Alerts for formal notice to industry regarding security issues:

- *Industry Advisory* - Purely informational, intended to alert registered entities to issues or potential problems. A response to NERC is not necessary.
- *Recommendation to Industry* - Recommends specific action be taken by registered entities. Requires a response from recipients as defined in the Alert.
- *Essential Action* - Identifies actions deemed to be "essential" to bulk power system reliability and requires NERC Board of Trustees approval prior to issuance. Like recommendations, essential actions require recipients to respond as defined in the Alert.

The risk to the bulk power system determines selection of the appropriate Alert notification level. Generally, NERC distributes Alerts broadly to users, owners, and operators of the bulk power system in North America utilizing its Compliance Registry. Entities registered with NERC are required to provide and maintain up-to-date compliance and cyber security contacts. NERC also distributes the Alerts beyond the users, owners and operators of the bulk power system, to include other electricity industry participants who need the information. Alerts may also be targeted to groups of entities based on their NERC-registered functions (e.g.; Balancing Authorities, Planning Authorities, Generation Owners, etc.)

Alerts are developed with the strong partnership of Federal technical organizations, including the Department of Homeland Security and the Department of Energy National Laboratories, and bulk power system subject matter experts, called the HYDRA team by NERC. NERC has issued 14 CIP-related Alerts since January 2010 (12 Industry Advisories and two Recommendations to Industry). Those Alerts covered items such as Aurora, Stuxnet, Night Dragon and the reporting of suspicious activity. Responses to Alerts and mitigation efforts are identified and tracked, with follow-up provided to individual owners and operators and key stakeholders. In addition, NERC released one Joint Product CIP Awareness Bulletin in collaboration with DOE, DHS and the FBI titled, "Remote Access Attacks: Advanced Attackers Compromise Virtual Private Networks (VPNs)".

The NERC Alert system is working well. It is known by industry, handles confidential information and does so in an expedited manner. The information needed to develop the Alert is managed in a confidential and expedited manner and does not require a NERC balloting process.

NERC understands that the Congress is seeking to ensure the cybersecurity of the electricity grid. Using standards, Alerts and essential actions, NERC is already working with FERC and the industry to protect the cybersecurity of the bulk power system.

NERC Work with DOD, DHS and DOE to Protect Grid Cybersecurity

As chair of the Electricity Sub-Sector Coordinating Council (ESCC), I work with industry CEOs and our partners within the government, including the Department of Defense, the Department of Homeland Security and the Department of Energy, to discuss and identify critical infrastructure protection concepts, processes and resources, as well as to facilitate information sharing about cyber vulnerabilities and threats. This type of public/private partnership is key to effective cybersecurity protection.

Recently, I met with officials from U.S. NORTHCOM where we discussed collaborating on various electric grid-focused activities including participation in the 2011 SecureGrid Exercise, providing electric sector situational awareness and collaborating on the Joint Capability Technology Demonstration (JCTD) Smart Power Infrastructure Demonstration for Energy Reliability and Security (SPIDERS). The latter project is being proposed to understand how specific facilities could develop small reliable "micro-grids" on a short-term or emergency basis. Similarly, NERC is discussing a project with DOD to develop case studies at critical military installations to further understand the requirements for "flow of power" and the implications to military readiness.

NERC is working with DHS National Cybersecurity and Communications Integration Center to develop a Memorandum of Understanding for bi-directional sharing of critical infrastructure protection information between the government and the electricity sector in North America. NERC also provides leadership to two significant DHS-affiliated public-private partnerships. These are the Partnership for Critical Infrastructure Security (PCIS) and the Industrial Control Systems Joint Working Group (ICSJWG). The PCIS is the senior-most policy coordination group between public and private sector organizations. On the government side, PCIS comprises the National Infrastructure Protection Plan (NIPP) Federal Senior Leadership Council (FSLC) and the State, Local, and Tribal Government Coordinating Council (SLTGCC), as well as the chairs of all of the other Government Sector Coordinating Councils. On the private side, PCIS comprises the chairs of all of the private-sector coordinating councils. The ICSJWG is a cross-sector industrial control systems working group that focuses on the areas of education, cross-sector strategic roadmap development, coordinated efforts on developing better vendor focus on security needs and cybersecurity policy issues.

NERC is engaged with DOE National Laboratories to further the level of awareness and expertise focused on cybersecurity, especially as it pertains to the bulk power system. We are working with Pacific Northwest National Laboratory on the Electric Sector Network Monitoring initiative and also on developing cybersecurity certification guidelines for Smart-Grid Cyber Operators. In a similar fashion, NERC is working with the Idaho National Laboratory to promote the Cyber Security Evaluation Tool for use within the electric sector. NERC also is partnering with the Industrial Control Systems Cyber Emergency Response Team to share threat, vulnerability and security incident information.

Finally, NERC is working with DOE and the National Institute of Standards and Technology to develop comprehensive cybersecurity risk management process guidelines for the entire electric grid, including both the bulk power system and distribution systems. We believe this to be particularly important with the increasing availability of smart-grid and smart-meter technologies. While the majority of technology associated with the smart grid is found within the distribution system, vulnerabilities realized within the distribution system could potentially impact the bulk power system. Everyone engaged in smart-grid and smart-meter implementation should ensure that appropriate security applications and technologies are built into the system to prevent the creation of additional threats and vulnerabilities.

NERC Comments on the Discussion Draft

First and foremost, NERC has consistently supported legislation authorizing some government entity to address cyber emergencies, as the draft would authorize the Secretary of Energy to do.

Second, NERC strongly supports any effort to improve information sharing between government and the private sector owners of critical electric infrastructure. NERC especially commends the provisions of the discussion draft directing the Secretary and the Commission to establish procedures on the release of critical infrastructure information to entities subject to the proposed legislation. NERC and the electric industry can only deal with the risks they are aware of. It is impractical, inefficient and impossible to defend against all possible threats or vulnerabilities. Entities must prioritize their resources to ensure they are protected against those risks that pose the greatest harm to their assets, their business and their customers. The electric industry is in the best position to understand the impact that a particular event or incident could have on the bulk power system, but the industry does not have the same access to actionable intelligence and analysis that the government does. This lack of information leads the industry to be, at best, a step behind when it comes to protecting against potential threats and vulnerabilities. Too often the industry has heard from government agencies that the threats are real, but is given little or no additional information. This leads to frustration among the private sector leaders who are unable to respond effectively due to ill-defined and nebulous threat information.

NERC also appreciates the additional attention in the discussion draft to providing security clearances, but that route will not likely deal with the unavailability of actionable information for electricity industry decision-makers. NERC has over 1900 entities on its Compliance Registry, some have just a few employees and some have many thousands. It is important to be realistic about the number of clearances that may be made available. Of more importance is developing methods and procedures for sanitizing sensitive information so that it can usefully be made available to the broad range of private decision-makers who must take action to protect against the threat or vulnerability.

The bulk of NERC's comments are directed to the draft legislation's treatment of "Cyber Security Vulnerabilities," which are something less urgent than "Cyber Security Threats." NERC appreciates that the draft legislation proposes for the ERO to play a meaningful role in addressing cybersecurity vulnerabilities, as the ERO now does. As discussed above, NERC has the tools, the expertise and the relationships with government agencies, intelligence resources and industry subject matter experts to address identified vulnerabilities effectively and efficiently. FERC has the authority now under FPA Sec. 215(d)(5) to direct NERC to prepare a proposed standard to address a specific vulnerability or other matter, and to do so by a certain date. Thus, it is not clear to NERC that the vulnerability section (proposed new FPA Section 224(b)) is needed. If this section is retained, please consider the following concerns:

1. *FERC's jurisdiction under this bill extends to distribution systems; the ERO's does not*: The definition of Critical Electric Infrastructure in proposed Section 224 extends to distribution systems. Section 215 does not provide NERC with that jurisdiction. Thus, existing NERC reliability standards and requirements cannot be as broad as FERC's jurisdiction under the draft bill, and standards prepared by NERC at the direction of FERC similarly cannot be as broad as FERC's direction if FERC directs an action to protect the distribution system action. If NERC is intended to have the same jurisdiction as FERC over the distribution system and assets, this needs to be clarified. Without such clarification, FERC could always find that an ERO-proposed reliability standard "fails to provide adequate protection of critical electric infrastructure from a cybersecurity vulnerability" and reject the ERO's efforts under Section 224, effectively removing the ERO role from the vulnerabilities section.
2. *Identification of vulnerability:* No requirement exists in the legislation for FERC to identify any deficiency in existing reliability standards or the specific cybersecurity vulnerability for the ERO to address. Without some idea of the "target" that FERC would like the ERO to hit, it will be difficult for the ERO to produce an adequate set of requirements, assuming the jurisdiction issue above is addressed.
3. *Enforceable tools in addition to standards:* The discussion draft calls for the ERO to develop a reliability standard in response to a FERC order on vulnerabilities, but given the constantly changing nature of vulnerabilities, not all vulnerabilities can or should be addressed by a

standard. Currently, NERC actions other than standards are not legally enforceable. Legislation that provides a means for both standards and other NERC directives to be legally enforceable would significantly enhance the cybersecurity of the grid. Such an approach would require the involvement of both the ERO and the Commission.

4. *Due process:* The discussion draft would authorize FERC to promulgate an interim final rule without consultation or any due process. In addition, unlike the 90-day sunset on DOE emergency orders, there is no such limitation on FERC interim final rules.

CONCLUSION

NERC works with multiple agencies, industry, consumers and government to support a coordinated comprehensive effort to address cybersecurity. As outlined today, NERC has many tools available including the ESCC and the ES-ISAC to address imminent and non-imminent threats and vulnerabilities through our Alerts and standards processes. These existing processes should be enhanced, not pre-empted, by cybersecurity grid legislation.

We appreciate this opportunity to discuss NERC's activities on cybersecurity with the committee and to offer our views on legislation that would improve cybersecurity protection of the grid.

End Note

[1] The Bulk Power System (sometimes referred to as "BPS") is defined as generation and transmission of electricity greater than 100kv, in contrast to the distribution of electricity to homes and businesses at lower voltages.

INDEX

D

E

F

G

H

I

S

T

U

V

W